Arnd BERNAERTS

Arctic Heats Up

Spitsbergen 1919 -1939

Black & White

Version

iUniverse books may be ordered through booksellers or by contacting:

iUniverse
1663 Liberty Drive
Bloomington, IN 47403
www.iuniverse.com
1-800-Authors (1-800-288-4677)

ISBN: 978-1-4401-4087-7 (sc)
ISBN: 978-1-4401-4088-4 (ebook)

Printed in the United States of America

iUniverse rev. date: 7/16/2009

1919

Spitsbergen:
***P**robably the greatest yet known*
temperature rise on earth.
B.J. Birkeland, 1930

1938

***I**n recent years attention is being directed more*
and more towards a problem which may possibly prove
of great significance in human affairs, the rise of temperatures
in the northern hemisphere, and especially in the arctic regions.
C.E.P. Brooks, 1938.

2009

***T**he ice in the polar region is melting away at an unprecedented speed during the last decade. Is the Arctic screaming? Is a tipping point reached? Is the ice melting apocalyptic? Indeed, during the last 150 years the sea ice retreat has never been as severe as in recent years. But, what is the cause? Was the current warming set in motion by the early Arctic warming which started since the late 1910s, and lasted from 1919 to 1939? The book will provide astonishing, but important answers.*

NOTE:
A colored edition has been given in print in Europe under the title:
How Spitsbergen Heats the World
The Arctic Warming 1919-1939

Contents

As the figures have been prepared in color their presentation here in black & white may not always indicate the full meaning; therefore kindly note that:
(1) the figures and graphs presented are available in color and enlargeable at: www.arctic-heats-up.com
(2) that due to techical reasons, for the b/w edition few figures needed to be relocated and/or merged into one image.

- Changes and modification of book/web text and figures reserved-

Chapter 1
Reviewing the past to understand the future -An Introduction

A. Arctic Warming – What Warming?

The claim that the summer of 2007 was apocalyptic for Arctic sea ice has recently gone around the globe, because the coverage and thickness of the sea ice in the Arctic has been declining steadily over the past few decades[1]. For many scientists this situation appears to be related to global warming (Brönnimann, 2008). In 2003 a USA research center formulated it this way already: "Recent warming of Arctic may affect worldwide climate"[2] Not everyone agreed but quarrel: What Arctic Warming?[3]

Although there is hardly a convincing reason to neglect the recent warming in the Arctic and the extent of ice melt during the summer season, it is not necessarily clear yet, whether the current discussion is based on a sound and comprehensive assessment. Climate research should not only deal with Arctic warming based on observations made during the last few decades, but at least be extremely interested in other climatic events that occurred in modern times, especially if somehow in connection with the situation in the Arctic. Why?

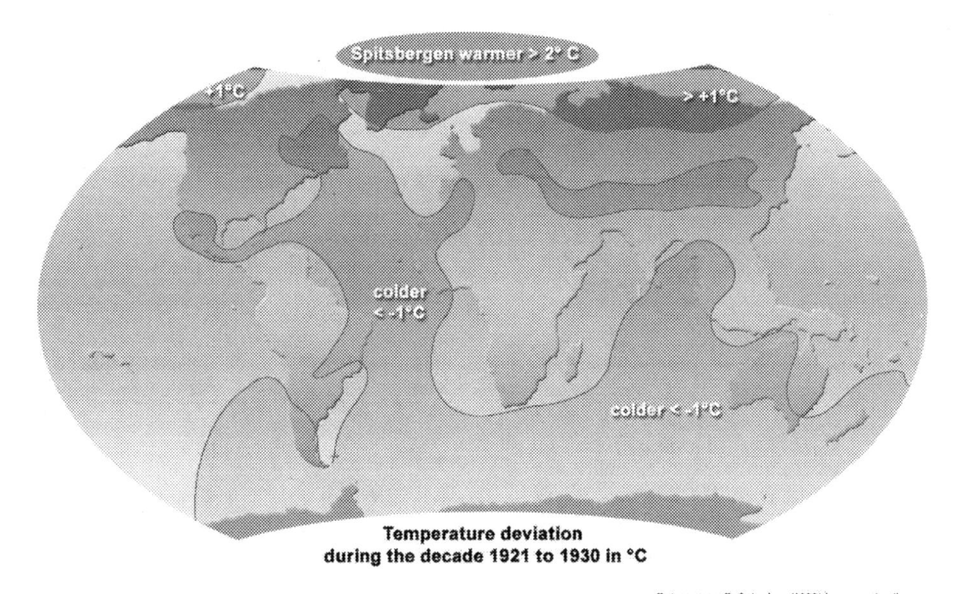

[1] Realclimate (web log); david: "Arctic sea ice: is it tipped yet", the 13th of December 2007. http://www.realclimate.org/index.php
[2] NASA; 23 Oct. 2003; http://www.nasa.gov/centers/goddard/news/topstory/2003/1023esuice.html
[3] Milloy, Steve, 13 Oct. 2005; http://www.foxnews.com/story/0,2933,172188,00.html

2008 Rune Graversen et al.

Extract from: http://www.arctic-warming.com, 7 January & 22 April 2008

Graversen et al., 2008, on: Vertical structure of recent Arctic warming

Rune G. Graversen, T.Mauritsen, M.Tjernström, E.Källén, G.Svensson; Nature, 3 January 2008, 451, p. 53-56

How much contributes this study on the "structure of recent Arctic warming" to understand the 'climatic revolution' (Ahlman, 1946) during the first half of last Century? Rune G. Graversen et al.'s article in the first 2008 issue of NATURE[1] got immediate attention world wide. The authors conclude: "We regress the Arctic temperature field on the atmospheric energy transport into the Arctic and find that, in the summer half-year, a significant proportion of the vertical structure of warming can be explained by changes in this variable. We conclude that changes in atmospheric heat transport may be an important cause of the recent Arctic temperature amplification." Some understood this immediately as confirmation that nature is pushing the Arctic to the edge, too. The study confirms according Seth Borenstein (AP[2]) that "There's is a natural cause that may account for much of the warming". 'Climate Feedback'[3] disagreed: Graversen conclusion only means: "Changes in the circulation in the atmosphere might have had a much larger effect than previously thought, but these changes may also have been induced by greenhouse gases". Does the explanation explain anything? Already back in the year 1938 C.E.P. Brooks asked: to account for the change in circulation.

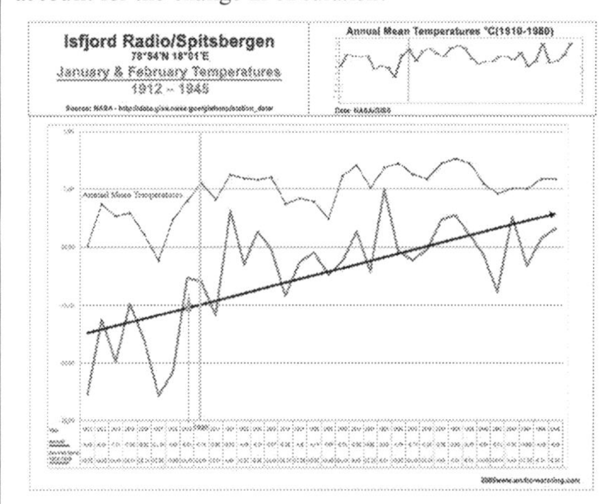

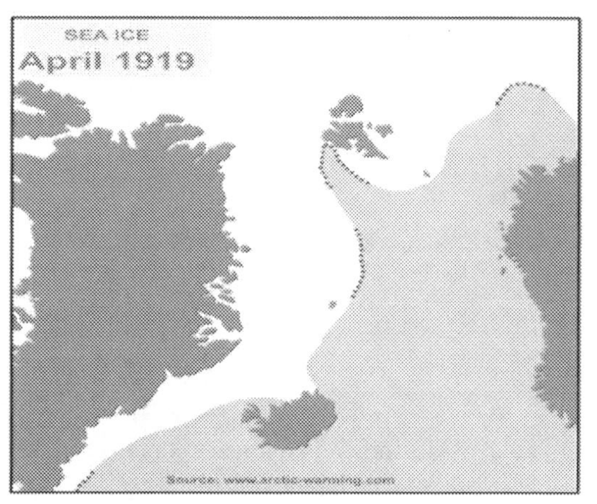

"Water is the driver of nature", *Leonardo da Vinci (1452-1519)*

Theses from Stockholm University, 2008, R. G. Graversen: „On the recent Arctic Warming"

Extract from: http://www.arctic-warming.com, 22 April 2008

Rune Grand Graversen' Doctor Thesis assumes[4] that: „A major topic is the linkage between the midlatitude circulation and the Arctic warming. It is suggested that the atmospheric meridional energy transport is an efficient indicator of this linkage". When Graversen concludes that the snow and ice-albedo feedbacks are a contributing but not dominating mechanism behind the Arctic amplification, and that a coupled climatemodel experiment with a doubling of the atmospheric CO2 concentration reveals a considerable Arctic surface-airtemperature amplification in a world without surface-albedo feedback, one is left to wonder, why such a thesis ignores completely the extreme winter warming from 1918 to 1922 which lasted until 1940. In this scenario CO2 is presumably the weakest mean to influence surface temperatures, and climate modeling is hardly a helpful tool, as long as not more distinctions between the sunless winter season and summer time is made.

---

[1] Rune G. Graversen, T.Mauritsen, M.Tjernström, E.Källén, G.Svensson; Nature, 3 January 2008, 451, p. 53-56
[2] Pioneer Press, 02 Jan. 2008; Nature giving global warming a nudge in Arctic, scientist says
[3] http://blogs.nature.com/climatefeedback/2008/01/arctic_amplification_1.html#more
[4] Theses from Stockholm University; http://www.divaportal.org/su/theses/abstract.xsql?dbid=7473.

1. Reviewing the past to understand the future – An Introduction

On the 2nd of November 1922, The Washington Post published the following story: Arctic Ocean Getting Warm; Seals Vanish and Icebergs Melt". The corresponding report in the Monthly Weather Review of November 1922 had also stated that the ice conditions in the Northern North Atlantic were exceptional; in fact, so little ice has never before been noted[4]. Only 16 year later the meteorologist C.E.P. Books thought it necessary to explain the situation more complex:

> *In recent years attention is being directed more and more towards a problem which may possibly prove of great significance in human affairs, the rise of temperature in the northern hemisphere, and especially in the Arctic regions.* (Brooks, 1938)

At the time of the writing of these lines in 1938, the Arctic had got as warm as in the first decade of the 21st Century. How much do we know about the mechanism that caused the previous arctic warming? Not very much, as Brönnimann et a. acknowledged: *"Our understanding of the climate mechanism operating in the Arctic on different timescales is still limited"* (Brönnimann, 2008). Is it reasonable and fair to dramatize the shrinking sea ice during a recent time period, if one is not fully aware of what happened in the early years of the last century?

Before the next chapter, we will insist on the question: why climatology should be able to explain the earlier arctic warming. Some phrases currently used are briefly presented in order to keep a context between the two warming periods, although this book primarily deals with the warming that The Washington Post reported already in 1922. Because the Where, When, and Why are still quite open, and by far not settled. One could actually describe the purpose of the book to answer a question that V.F. Zakharov (1997) submitted a decade ago:

- *Why are the maximum climate fluctuations confined to the Atlantic sector of the Arctic?*
- *Why are these fluctuations pronounced, first of all, right here?*
- *Should the Atlantic sector of the Arctic be considered as a center of some kind, a source of climate change over the Hemisphere?*

The focus is clear: What role did the ocean play? The investigation will prove that it had been substantial, by time, intensity and duration. But once these aspects have been thoroughly elaborated, the discussion will be extended to the question: Why? After all, the first arctic warming began at the end of the World War One in the winter of 1918-19, and died away when the Second World War began on the 1st of September 1939. That is worth a discussion, even if it is not the purpose of this paper to offer conclusive evidence in this respect

B. The Arctic Is Screaming?

The world should know: The Arctic is Screaming noted newspapers recently. No one had heard the Arctic crying, but was there something that should have signaled horror? It is true; the annual arctic sea ice cover had been decreasing during the summer season for a couple of years. The remaining minimum ice cover around September produced record after record: the record from 2005 was beaten by 2006, which was beaten by 2007. That was the point when the émigré in polar science Mark Serreze informed the press:

---

[4] Ifft; George N., 1922, „The Changing Arctic", Monthly Weather Review, Nov 1922,

1919

Arctic Warming started – 90 years ago – January 1919 - Does the Arctic scream, and why?

By http://www.oceanclimate.de, 10 January 2009

Suddenly, in January 1919, Spitsbergen a remote archipelagos between the North Cape in Norway and the North Pole had corresponding mean temperatures during January as 2000 km further south in Oslo, mere –5°C. That was a climatic 'bang'. It meant that the temperature differences between the two pre and post WWI January was 16° (sixteen) degrees[1]. The extraordinary situation during this month went unnoticed. The event itself not.

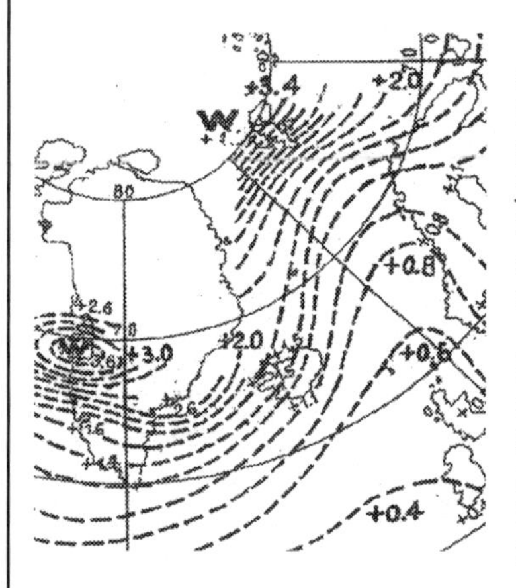

Soon the Arctic started to warm. In 1930 the Norwegian scientist B. J. Birkeland[2] regarded the rise as possibly the biggest ever observed at a single place. Presumably Birkeland could repeat the claim today again with justification. In 1932 the New York Times reported: " Next great deluge forecast by scientists – Melting polar ice caps to raise the level of the seas and flood the continents"[3]. Only three decades after January 1919 the developments in the Arctic had been called: a climatic revolution[4].

How dramatic the rise was is possibly best illustrated with few graphics, which indicate the extraordinary temperature jump. Particularly explanatory is the image showing the global situation in the years after the event from 1921 –1930, and how the temperature situation effected the Arctic and Europe toward the end of the 1930s, just before the warming ended and a three decade long cooling phase stated with WWII in winter 1939/40.

<u>Spitsbergen with a plus of 3.4° C in winter 1921-1930</u>

The question today is, what do we know about the January 1919 Spitsbergen event? Do we have any information whether the early Arctic warming from 1919 to 1940 has been in any way a source of the dramatic sea ice melting in the Arctic over the last few years? Unfortunately not. Science has shown little interest to provide explanation. Instead we are told that this is "one of the most puzzling climate anomalies of the 20th century"[5].

Neither helpful is this explanation either: "The recent dramatic loss of Arctic sea ice appears to be due to a combination of a global warming signal and fortuitous phasing of intrinsic climate patterns"[6] Recently a scientist said: "The Arctic is screaming"[7]. Presumably the Arctic is wondering why the Spitsbergen event in January 1919 and subsequent warming is not understood and explained after 90 years.

Graphic: Cut-out from 'Tafel 58'; Decade 1921-1930; Deviation of winter temperature (Nov-March) from long-term mean; R. Scherhag, (1936/Sept); Ann. Hydrologie & Maritimen Meteorologie, Sept. 1936, p. 397-407.

---

[1] The January temperature at Spitsbergen : 1917 -20.4; 1918-24. 4; (mean: 22,4°C); and 1919 -5.7, 1920 -10.5 (mean: -8,1°C); http://data.giss.nasa gov/work/gist emp/STATIONS//tmp.634010050010. 1.1/station.txt

[2] Birkeland, B.J. (1930) , Temperaturvariationen auf Spitzbergen, Meteorologische Zeitschrift, Juni 1930, p. 234-236.

[3] New York Times; May 15, 1932.

[4] Ahlmann, H.W. (1946); "Research on Snow and Ice, 1918-1940", The Geographical Journal, 1946, p.11-25

[5] Bengtsson, et al (2004), The Early Twentieth-Century Warming in the Arctic—A Possible Mechanism, Joumal of Climate, page 4045-4057.

[6] Overland, J. E. (2008); M. WANG & S. SALO; „The recent Arctic warm period", Tellus, 2008, 60A, p.589–597

[7] This was widely reported, e.g. Associated Press; 12 Dec.2007 by Seth Borenstein; "Ominous Arctic Melt Worries Experts".

1. Reviewing the past to understand the future – An Introduction

"The Arctic is screaming"[5]. As a senior scientist at the government's snow and ice data Center in Boulder, Colorado, Mark Serreze should know what he is talking about, or had it been his scream? At least the time for setting a scream was timely. Special legislation for polar bears was already on the way, as it was assumed that the dwindling sea ice would not leave the bears without ice floats but affecting wildlife widely.

"Greenland's ice sheet melted nearly 19 billion tons more than the previous high mark, and the volume of Arctic sea ice at summer's end was half what it was just four years earlier[6]. At this rate, the Arctic Ocean could be nearly ice-free at the end of summer by 2012, much faster than previous predictions", the NASA climate scientist Jay Zwally was cited. But even this prediction could be topped by arctic experts claim to a conference meeting that, if the Arctic sea ice was melting so rapidly, as it recently did, than any sea ice throughout the Arctic Ocean could have entirely disappeared by the summer of 2040.

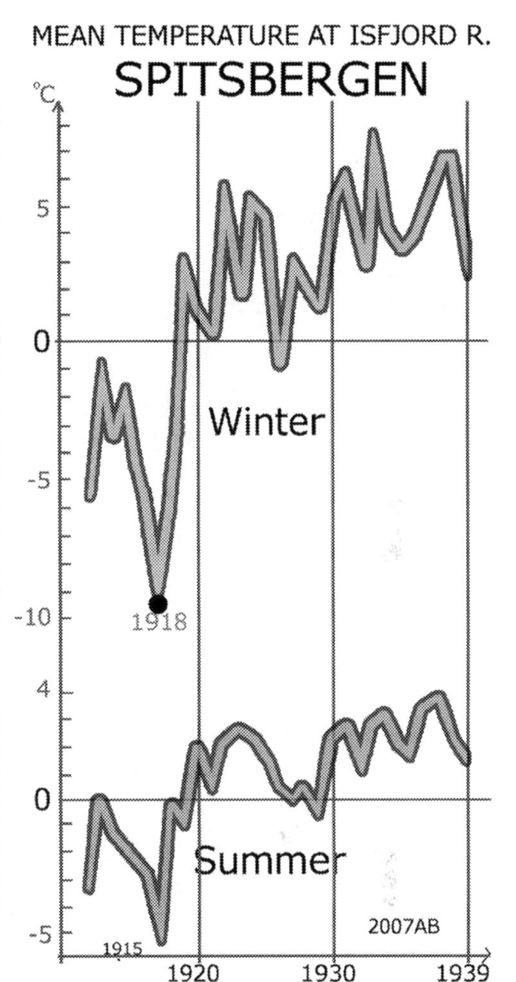

Image based on source: Hesselberg and Johannessen, 1958

In this context, the CNN could observe that scientists have been asking themselves these questions: Was the record melt seen all over the Arctic in 2007 a blip amid relentless and steady warming? Or has everything sped up to a new climate cycle that goes beyond the worst case scenarios presented by computer models? Nobody has given any answer. But Mark Serreze says: the Arctic is screaming. Is it impossible to find out?

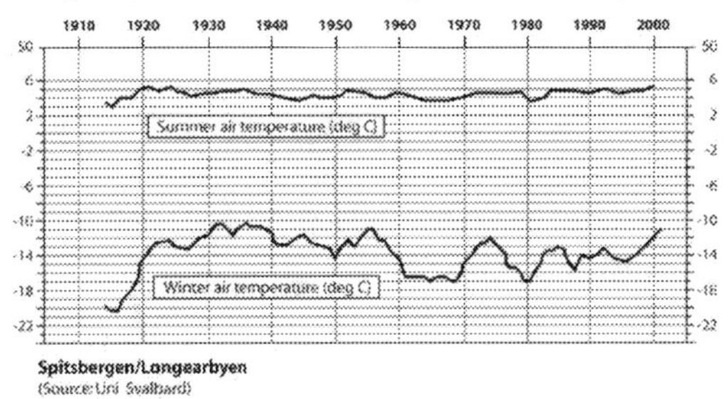

It s hould not be impossible, as the present arctic warming since 1980 is not the only one. There was another warming period for the region north of 62° North since 1920 until 1945, for which the high-latitude temperature increase was stronger in the late 1930s early 1940s than in recent decades (Polyakov, 2002). The first Arctic warming started 90 years ago, from about 1920 to 1940[7]. In winter 1918-19 the air temperatures exploded at the remote archipelagos Spitsbergen, which the Norwegian call: Svalbard. In 1930 the Norwegian

[5] This was widely reported, e.g. Associated Press; 12 Dec.2007 by Seth Borenstein; "Ominous Arctic Melt Worries Experts".

[6] CNN; 11th of December 2007: http://www.cnn.com/2007/TECH/science/12/11/arctic.melt.ap/index.html

[7] M. Serreze (Serreze, 2006) acknowledges that *"Substantial high-latitude warming from about 1920 to 1940 was followed by cooling until about 1970, then another period of marked warming that extends through the present"*, but makes little effort to understand the early warming in the first place, but provides as conclusion: "One important piece of evidence supporting an enhanced GHG (Green House Gas) contribution is that while the earlier 20th century warming is only seen at higher latitudes, indicative of natural variability in the climate system, the recent warming is apparent in all latitude zones." Concerning 'the piece of evidence' Serreze sees now, a couple of years ago he and some colleagues claimed the opposite (Kahl, 1993).

scientist .B. J. Birkeland [8]regarded the rise as maybe the biggest ever observed in one place. Birkeland could presumably repeat the claim today with justification. But would he or his colleagues come up with the exclamation the Arctic is screaming today? Definitely not, although he and his colleagues might wish to scream: How could it be that you know so little about "our arctic warming" to understand "your arctic warming".

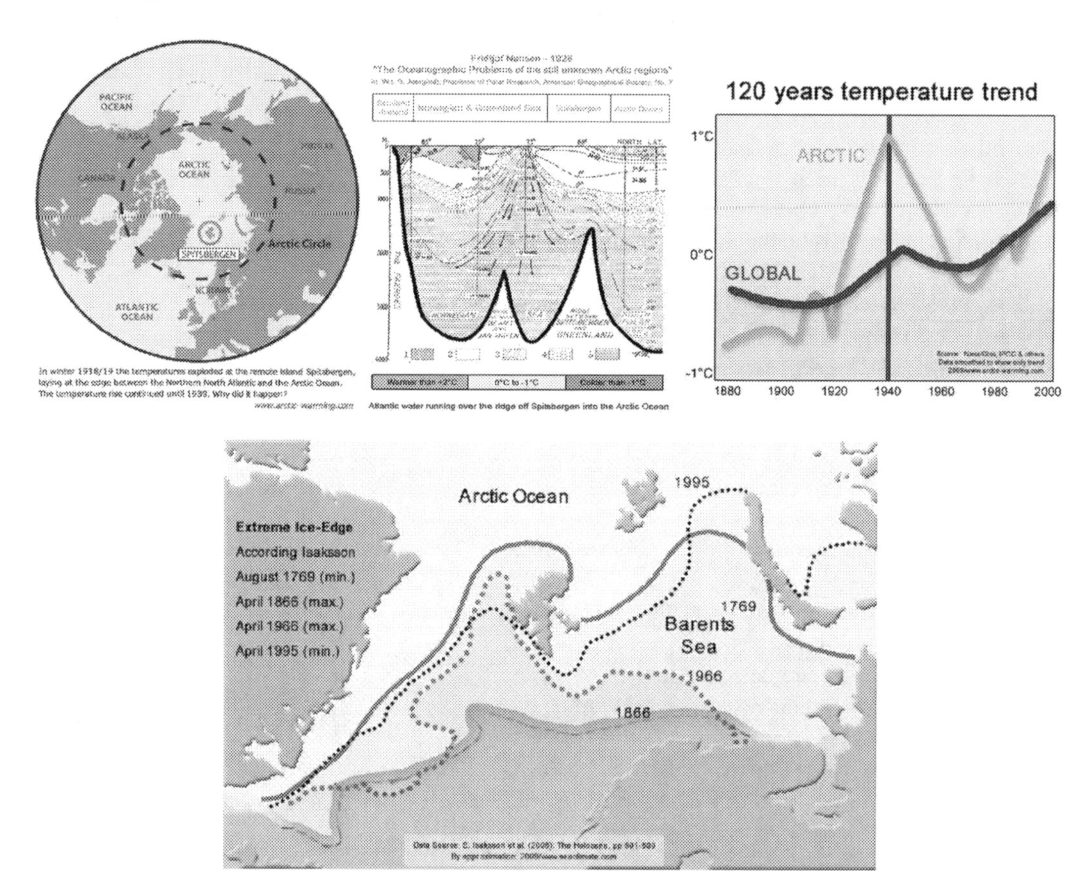

C. Tipping Point, Climatic Revolution, Apocalypse Now

Few years ago scientists called for "cutting emissions now to avoid climate tipping point"[9]. The phrase has reached the Arctic. The phrase actually refers to the fact that in physics, if a small amount of weight is added to a balanced object, it can topple the balance completely and very suddenly. After the ice season in September 2007 had been the lowest since the Little Ice Age around 1850. Is the decreasing seasonal sea ice extent during the summer season the tipping point? At least it could be the imminent example of a tipping point in the climate system that has been argued[10].

---

[8] Birkeland, B.J. (1930) , Temperaturvariationen auf Spitzbergen, Meteorologische Zeitschrift, Juni 1930, p. 234-236.
[9] The Guardian/UK, 2005, Ian Sample, "Warming hits 'tipping point", Thursday, the 11th of August 2005
[10] Realclimate (web log); David: "Arctic sea ice: is it tipped yet", the 13th of December 2007.

1928

Fridjof Nansen – 1928

The Oceanographic Problems of the Still unknown Arctic Region[1]

Extract from section: Circulation of the Water

A methodical study of the water layers and their movements in the still unknown regions of the North Polar Sea will be of much interest. As was discovered during the FRAM expedition of 1893 – 1896, this sea is covered by a layer, 150 to 200 meters thick, of cold water with temperatures between 0°C and –1,9°C and a comparatively low salinity owing to the admixture of fresh water, chiefly river water from Siberia, Alaska, and Canada. Below this surface layer there is a layer , some 600 to 700 meters thick, of warmer and salter water, with temperatures above 0°C. and salinities approaching 35 per mille. This is Atlantic water which is carried into the Arctic Basin chiefly by the small branch of the Atlantic Current ("Gulf Stream") running northwards along the west coast of Spitsbergen. Below this warmer water there is again colder water filling probably the whole basin to the bottom; its temperatures between 0°C. and –0,8°C and its salinity 34.90 per mille. This cold deep-water originates in the northern part of the Norwegian Sea, north-northeast of Jan Mayen, where it sinks down from the surface, which is cooled by the radiation of heat during the winter and spring. The thus cooled water runs into the Arctic Basin across the probable submarine ridge between Spitsbergen and Greenland. A study of the condition of these various water layers and their distribution in the various parts of the North Polar Sea would be of much value. While we drifted with the FRAM across the Arctic Basin our deep-sea observations showed that the boundaries between the water layers, especially between the cold surface layer and the warmer underlying water, were subjected to considerable vertical oscillations. By later observations we have found that such vertical oscillations, due to surface boundary waves, often of very considerable dimensions, probably are quite common phenomena in the ocean; but they have not yet been sufficiently studied methodically. From the drifting ice movements of the water – the horizontal currents as well as these vertical oscillations of the layers – may be continually and carefully studied at all depths in an ideal manner which is not possible in the open ocean; and many of the greatest problems of oceanography may thus be solved.

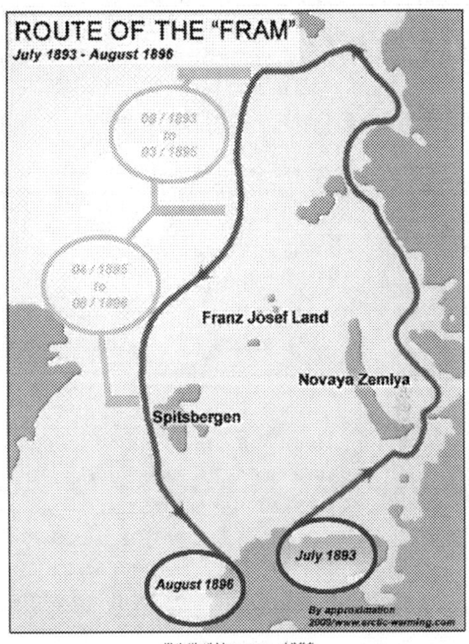

Fridjof Nansen - 1928
"The Oceanographic Problems of the still unknown Arctic regions"
in: W.L.G. Joerg(ed); Problems of Polar Research, American Geographical Society, No. 7

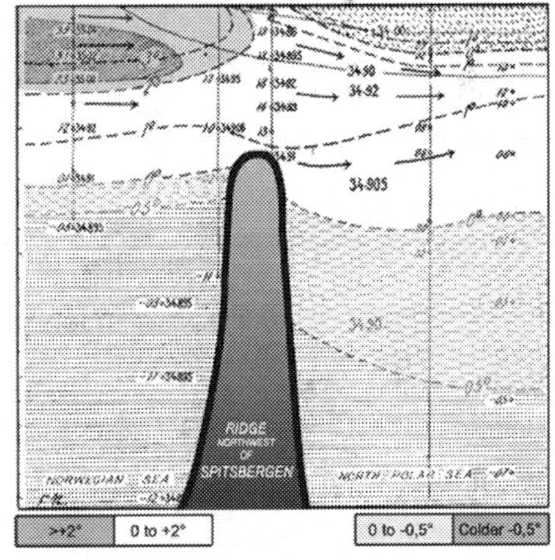

Atlantic water running over the ridge off Spitsbergen into the Arctic Ocean

[1] Nansen, Fridjof (1928);in Joerg, W.LG.;(ed): Problems of Polar Research, American Geographical Society, Special Publication No. 7, p.3 ff

1. Reviewing the past to understand the future – An Introduction

The Russian Revolution took place 90 years ago. At a tipping point communism replaced the monarchy. Only one year later a pronounced warming started in the Atlantic-Arctic region, lasting for two decades. Its impact on temperatures, sea ice and glaciers had been so significant that the Norwegian scientist Ahlmann estimated that this warming period had been a 'climatic revolution' (Ahlmann, 1946). Which of the phrases is more serious? We do not know? We know too little about the earlier warming. We do not know whether the earlier warming of the Arctic is a precondition for the situation in the Arctic now. In one way or another, it will be the case, and therefore the identification of any tipping point in the previous arctic warming is as much required as talking about the recent situation.

Also the NASA scientist James Hansen suggests that the Earth may have hit the tipping point, which, according to his interpretation, means: "We have not passed a point of no return. We can still roll things back in time – but it is going to require a quick turn in direction"[11]. To Hansen, who is regarded as the godfather of global warming since he presented the subject to the US Senate in 1988: "the 'point of no return' is in being, when climate reaches a point with unstoppable irreversible climate impacts (irreversible on a practical time scale)" as he explained recently at a presentation[12] that Hansen exemplifies with the disintegration of large ice sheet in the Arctic.

Is the "Tipping Point" in the Arctic? Not necessarily, but according to some commentators' view after saying: "The summer of 2007 was apocalyptic for Arctic sea ice". It was not for the first time that the phrase had been used in climate research, but it is of a recent date. In 2005 a newspaper story was entitled: "Apocalypse Now: How Mankind is Sleepwalking to the End of the Earth Floods, storms and droughts. Melting Arctic ice, shrinking glaciers, oceans turning to acid. The world's top scientists warned last week that dangerous climate change is taking place today, not the day after tomorrow. You don't believe it?"[13].

Although the 'climatic revolution' is on record for more than 80 years, the current knowledge on: Where, When, and Why, is rather limited, as recently acknowledged that *"one of the most puzzling climate anomalies of the 20th century" (Bengtsson, 2004)*, while other offer merely the conclusion *"that the earlier warming was natural internal climate-system variability"* (Johannessen, 2004).

That seems too little and too superficial to deal with an unprecedented 'climatic revolution'. Much worse is the behavior of Oscar price winner Al Gore with his claim that men have put so much carbon dioxide in the thin shell of air surrounding the world that literally the Earth heat has been changed, causing a universal threat of cosmic in scale[14]. Back in the late 1910s men released only a small amount of greenhouse gases into the atmosphere. Here is certainly not the place to challenge the greenhouse thesis, respectively assessing its possible impact, but discussing the climate change issue in Al Gore's way seems irresponsible as long as a 'climatic revolution' that occurred under the eyes of modern meteorology has not thoroughly analyzed and the causation has been convincingly explained.

To put it clear, raising any reasonable concern on matters affecting human welfare, whether by tipping point or apocalyptic images is one thing, but sleepwalking the profound arctic warming during the early years of the last century is irresponsible. As long as this event is not well understood, any apocalyptic or tipping point talking is reckless and not very helpful.

---

[11] The Associated Press: Ominous Arctic Melt Worries Experts, by Seth Borenstein, the 12th of December 2007.
[12] Hansen in his: Bjerknes Lecture, to the American Geophysical Union, San Francisco, the 17th of December 2008; PowerPoint slide 26.
[13] Independent/UK, the 6th of February 2005; by Geoffrey Lean
[14] Al Gore, Moving Beyond Kyoto, The New York Times, July 1, 2007, WK 13.

1990

J.T.Houghton, et al. (ed), 1990, Climate Change
The IPCC Scientific Assessment, Chapter 7, Observed Climate Variation and Change, Cambridge 1990, p. 223

Even the upper few meters of the ocean can store as much heat as the entire overlying atmospheric column of air. Scientists have long recognized that the ocean could act to store large amounts of heat, through small temperature changes in its sub-surface layers, for hundreds or thousand of years. When this heat returns to the atmosphere/cryosphere system it could also significantly affect climate.

The magnitude and extent of the observed changes in the temperature and salinity of the deep North Atlantic are thus large enough that they cannot be neglected in future theories of climate change.

2007

The Arctic in the Eyes of the Intergovernmental Panel on Climate Change
Fourth Assessment Report 2007 (Abstract; References not shown)

IPCC; Summary for Policymakers (p.7)
Average arctic temperatures increased at almost twice the global average rate in the past 100 years. Arctic temperatures have high decadal variability, and a warm period was also observed from 1925 to 1945.

Chapter 5; Observations: Oceanic Climate Change and Sea Level
5.3.2.2 Arctic Ocean

Climate change in the Arctic Ocean and Nordic Seas is closely linked to the North Atlantic sub polar gyre. Within the Arctic Ocean and Nordic Seas, surface temperature has increased since the mid-1980s and continues to increase. In the Atlantic waters entering the Nordic Seas, a temperature increase in the late 1980s and early 1990s has been associated with the transition in the 1980s towards more positive NAO states. Warm Atlantic waters have also been observed to enter the Arctic as pulses via Fram Strait and then along the slope to the Laptev Sea; the increased heat content and increased transport in the pulses both contribute to net warming of the arctic waters. Multi-decadal variability in the temperature of the Atlantic Water core affecting the top 400 m in the Arctic Ocean has been documented. Within the Arctic, salinity increased in the upper layers of the Amundsen and Makarov Basins, while salinity of the upper layers in the Canada Basin decreased. Compared to the 1980s, the area of upper waters of Pacific origin has decreased. During the 1990s, changed winds caused eastward redirection of river runoff from the Laptev Sea (Lena River, etc.), reducing the low-salinity surface layer in the central Arctic Ocean, thus allowing greater convection and heat transport into the surface arctic layer from the more saline subsurface Atlantic layer. Thereafter, however, the stratification in the central Arctic (Amundsen Basin) increased and a low salinity mixed layer was again observed at the North Pole in 2001, possibly due to a circulation change that restored the river water input. Circulation variability that shifts the balance of fresh and saline surface waters in the Arctic, with associated changes in sea ice, might be associated with the NAM, however, the long-term decline in arctic sea ice cover appears to be independent of the NAM. While there is significant decadal variability in the Arctic Ocean, no systematic long-term trend in subsurface arctic waters has been identified.

1990

J.T.Houghton, et al. (ed), 1990, Climate Change
The IPCC Scientific Assessment, Chapter 7, Observed Climate Variation and Change, Cambridge 1990, p. 233

The rather rapid changes in global temperature seen around 1920 –1940 are very likely to have had a mainly natural origin.

2008

Missing the Point on Arctic Warming, Ø. Nordli, IPCC, NASA ?

By http://www.arctic-warming.com, August 2008

"The Arctic ocean is warming up, icebergs are growing scarcer and in some places the seals are finding the water too hot", reported the The Washington Post, on November 2nd, 1922. B.J. Birkeland (1930) saw the temperature rise, as *"probably be the greatest yet known on earth"*, and few years later A. W. Ahlmann (1946) called the event a 'climatic revolution'[1]. This site explains this sudden warming since winter 1918/19 in a detailed step-by-step approach(http://www.arctic-warming.com).

Since about the 1980th it is evident that the arctic is warming, after a colder period over four decade again. It is good that this trend receives attention since recently. IPCC has little problems to assert[2]: *The Arctic is expected to experience the greatest rates of warming compared with other world regions.* However, the early warming is not explained, and the little they say is inaccurate[3]. Other title it in this way: "NASA's Earth scientists think ice is hot - a hot topic, that is[4]", but fail to explaining anything either.

That is a big surprise as there are few, but reasonable data documented. Ø. Nordli, a scientist at the Norwegian Meteorological Institute, confirms the reliability of the data taken at Spitsbergen[5], stating: *An abrupt change of temperature occurred at the end of the 1910s transforming the Svalbard climate from a cold phase (1911-1919) to a warm phase (1920-1930).* Evidently Spitsbergen saw a temperature increase of more than 10 degrees Celsius from winter 1916 & 1918 to winter 1922/23. Despite this fact, Ø. Nordli made a statement concerning the period 1911 - 2004[6]: "During winter (DJF) no significant trend in the data is seen, whereas in spring the trend is highly significant, 0.42 °C per decade." It seems Ø. Nordli missed the most interesting and important point: What cause the temperatures to 'explode' in winter 1918/19?

Also IPCC is too superficial in this respect (see above). At least they should have paid attention to the advise, V.F. Zakharov submitted to the World Meteorology Organization (WMO) in 1997[7], asking:

(1) *Why are the maximum climate fluctuations confined to the Atlantic sector of the Arctic?";*
(2) *Why are these fluctuations pronounced, first of all, right here?";*
(3) *Should the Atlantic sector of the Arctic be considered as a center of some kind, a source of climate change over the Hemisphere?"*

Also a work from Sergey V. Pisarev (1997)[8] indicates that the impact of the sea may require more attention. Actually, this site is carefully elaborating the reasons for the sudden commencement of the arctic warming since winter 1918/19, concluding, that the source had been the seas around the Spitsbergen archipelagos as far as not covered by sea ice according the seasons. When Ø. Nordli observes: *"The cold phase was characterized by clear sky and pronounced inversions, whereas the warm phase was characterized by overcast sky and weaker and rarer inversions[9]"*, the answer is easy, for the winter season at least: It is the sea.

---

[1] See Chapter A, Introduction, http://www.arctic-warming.com/introduction-the-scope-of-this-investigation.php ,
[2] IPCC, 2007a: Climate Change 2007: The Physical Science Basis. Contribution of Working Group I to the Fourth Assessment Report of the Intergovernmental Panel on Climate Change [Solomon, S., D. Qin, M. Manning (eds.)].
[3] Ditto: Average Arctic temperatures increased at almost twice the global average rate in the past 100 years. Arctic temperatures have high decadal variability, and a warm period was also observed from 1925 to 1945.
[4] http://www.nasa.gov/vision/earth/environment/Arctic_Warming_ESU.html ;
[5] Øyvind Nordli, Year ?, "Temperature variations at Svalbard during the last century" at: http://www.nordicspace.net/PDF/NSA106.pdf
[6] Øyvind Nordli, 2005, „Long-term Temperature Trends and Variability at Svalbard (1911 – 2004)", Geophysical Research Abstracts, Vol. 7, 06939, 2005.
[7] Zakharov, V.F.; 1997, 'Sea Ice in the Climate System', Arctic Climate System Study, WMO/TD-No. 782, in the section "On the nature of 'polar forcing'", p. 71.
[8] Sergey V. Pisarev , 1997, "Arctic Warming" During 1920-40: A Brief Review of Old Russian Publications, http://mclean.ch/climate/Arctic_1920_40.htm
[9] Øyvind Nordli, 2005, „Long-term Temperature Trends and Variability at Svalbard (1911 – 2004)", Geophysical Research Abstracts, Vol. 7, 06939, 2005.

D. Natural Variables versus Anthropogenic

The last 100 years are marked by the changeover of a climate system dominated by natural forcing to a climate system dominated by anthropogenic influences (Brönnimann 2008). This shall be understood as attributing the warming trend before the 2nd World War to "natural variability" and the subsequent warming after the war to anthropogenic forcing, particularly due to the increasing release of carbon dioxide (CO2). This is so easily said and if we look thoroughly at every aspect, this may prove worthless.

For example, try to find a founded explanation for the meaning of 'natural climate variability", respectively corresponding phrases. The glossaries of the leading international organizations as WMO and UNFCCC secretariat are blank on this subject. The AMS however provides the following[15]:

> *climate variability* — temporal variations of the atmosphere– ocean system around a mean state. Typically, this term is used for timescales longer than those associated with synoptic weather events (i.e., months to millennia and longer). The term "natural climate variability" is further used to identify climate variations that are not attributable to or influenced by any activity related to humans.

But it can come along even more confusing[16]:

> The relatively short instrumental record of climate (the last 50 to 100 years), which reflects anthropogenic change as well as natural variations, does not represent a stationary or steady record. Instead, climate fluctuations over the past few millennia or so will need to be analyzed to establish a baseline of natural variability against which future (and present) variations can be gauged.

With reference to other work elsewhere that discusses the definition crux in meteorology in

SPITSBERGEN, 1912 TO 1926

Temperature deviation of monthly mean from a 15-year average

Year	Jan.	Febr.	Mar.	Apr.	May	Jun.	Jul.	Aug.	Sept.	Oct.	Nov.	Dec.	Yearly
1912	-8.4	-7.3	-3.4	-6.2	-1.2	-0.1	-1.3	-2.0	-2.9	-2.9	-1.2	+1.8	-3.1
1913	+0.3	-1.7	+0.7	+3.8	-0.4	-1.8	-0.8	+0.6	+0.6	-2.8	+4.0	+1.1	+0.2
1914	-5.7	-4.9	-1.4	+3.0	-0.4	-0.2	-0.5	+0.2	-1.0	+1.2	-1.5	-3.6	-1.3
1915	+1.8	-0.5	-3.1	+2.5	-4.0	-0.4	-1.5	-1.0	-0.6	+0.9	-9.0	-8.3	-2.0
1916	-8.6	+2.1	-2.3	-3.0	-1.2	+0.4	-0.4	-0.9	-0.2	-1.3	-5.9	-8.4	-2.5
1917	-7.4	-10.3	-8.7	-9.2	-4.8	-1.4	-2.1	-2.3	-2.8	-2.7	-2.6	-5.4	-5.0
1918	-10.1	-0.4	-0.1	+0.8	+2.4	+0.8	+1.9	-0.1	-0.4	-1.8	+0.9	+7.4	+0.1
1919	+8.6	-4.7	-6.9	-6.3	+3.3	+0.7	-1.1	+0.2	-0.6	+1.0	-4.0	+0.8	-0.8
1920	+3.8	+1.4	+8.9	+0.8	+1.9	-0.5	-0.5	+0.7	+0.4	+3.5	+4.3	+3.8	+2.3
1921	-0.8	+0.1	+2.3	+2.5	-0.2	+0.5	+1.0	+1.2	-0.9	-2.1	+1.3	+2.5	+0.6
1922	+10.5	+6.9	+0.1	-0.5	+1.1	+1.7	+2.1	+1.2	+1.4	+1.8	+5.0	-0.9	+2.5
1923	+3.3	+4.8	+5.9	+3.8	+2.3	-0.1	+1.4	+1.5	+1.5	+3.0	+3.9	+4.5	+2.9
1924	+5.7	+8.1	-1.9	+2.3	+2.6	-0.6	+2.1	+0.8	+0.9	+2.3	+3.1	+5.3	+2.5
1925	+4.3	+6.3	+7.7	+2.1	-0.9	+1.4	+0.1	+1.4	+3.1	-0.7	+1.7	-3.1	+1.9
1926	+2.2	+0.5	+1.5	+4.0	-0.8	-0.5	+0.1	-0.8	+1.5	+0.1	-0.7	+2.5	+0.8

Source: B.J. Birkeland, Meteorologische Zeitschrift, June 1930, p. 234

Winter temperature jump in 1919-1940

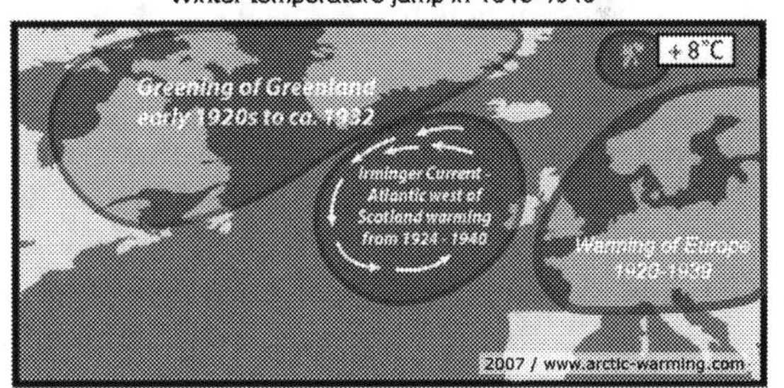

Winter Temperatures Deviation
1933 to 1935, November to March

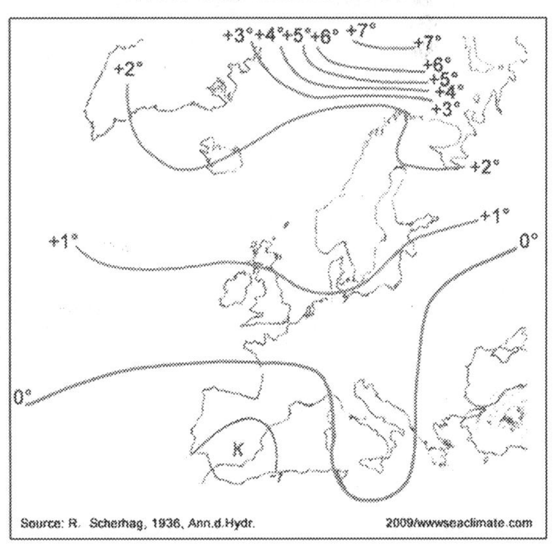

---

[15] Glossary of Meteorology, 2000, 2nd ed.; by the American Meteorological Society (AMS).

[16] CGER (1995) Commission on Geosciences, Environment and Resources; ""Natural Climate Variability on Decade-to-Century Time Scale"; http://books.nap.edu/openbook.php?record_id=5142&page=601

detail[17], the subject is circumvented here, to concentrate on the first arctic warming, but if deemed useful the matter will be raised again.

All this talking about 'natural variability' says nothing of substance about the mechanism and causation of the first arctic warming in modern times. It even makes little effort to ensure a full assessment and understanding about the earlier event. One may only recall a report by the New York Times in 1932 to realize what it could mean when the substance of claims are not clear: " Next great deluge forecast by scientists – Melting polar ice caps to raise the level of the seas and flood the continents"[18]. Is that very different from current claims? At least, C.E.P. Brooks (1938) had that to say with regard to this warming period: "There have been great climatic oscillations before this, even since the last Ice Age, about the causes of which we are quite ignorant". It seems little has changed when now confronted with the mere allegation, that the warming the NYT and Brooks are talking about has been due to natural variability, even without explaining which part of nature did something, and which did not.

But the phrase 'natural variability' seems not to be the end of terms which may cause more confusion than clarification as this example may illustrate: "The recent dramatic loss of Arctic sea ice appears to be due to a combination of a global warming signal and fortuitous phasing of intrinsic climate patterns" (Overland, 2008). How to make sense out of this?

The following investigation will show that the 1st Arctic warming was not understood in the 1930s, and that little has changed to better at the end of the first decade of the 21st century, about seven decades later. What is needed is a better understanding of what had happened almost a century ago in the high northern hemisphere, and this work does not agree with NASA expert Waleed Abdalati notion: "The first step in understanding why things happen is observing what is happening"[19]. It is the other way around. Without understanding the mechanism of the warming almost a century ago one might delay the understanding of the current ice melting. Even more, understanding the first arctic warming would definitely provide many hindsight for better assessing and handling the entire climate change issue.

---

[17] http://www.whatisclimate.com/

[18] New York Times; the 15th of May 1932

[19] Cited by: Stofer, Kati, (23rd of October 2003), "Seasons of Change: Evidence of Arctic Warming Grows", http://www.nasa.gov/vision/earth/environment/Arctic_Warming_ESU.html

Chapter 2
The Arctic in Climatology

A. The general picture

a) Remote but influential

The Arctic Ocean is rarely in focus of civil society due to its harsh environment and remote location. Even from an oceanic perspective it has not a big weight. Although it covers an area of about 14,056,000 km^2 (5,427,000 sq mi), which is almost the size of Russia, and has four basins with depth of 3000m to 4500m, its total volume is only 1.3% of the World Ocean. However this does not reflect the impact of this ocean in the climatic system, which is substantially higher than its size and water volume. The influence derives from a number of unique physical features, which exist and are at work only here. As main aspect one has at least to name:

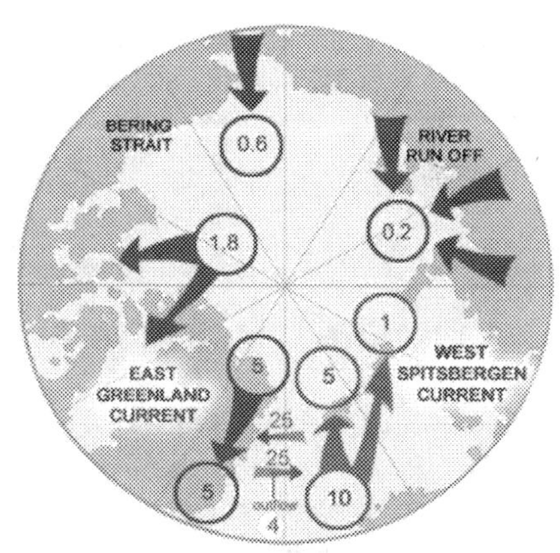

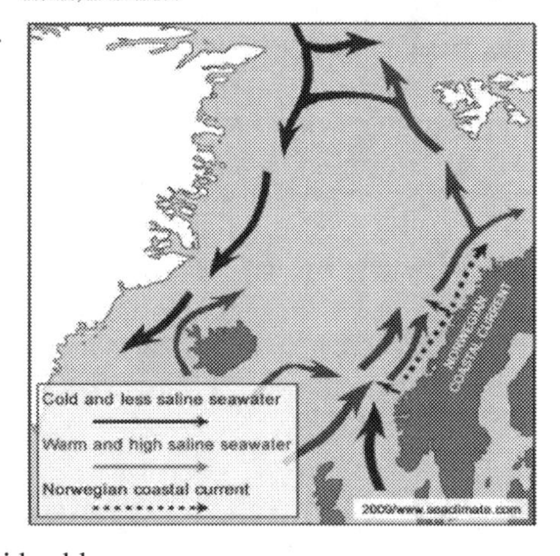

- The water body is very cold, generally slightly below 0°C (32 °F), albeit the water layer at the range from the surface to max. 200m has temperatures close to the freezing point (-1.5° to -1.9° C) with variations over the depth.
- In this extreme cold environment is the degree of salinity of extreme importance, as even minor changes in temperature, salinity and density leads to internal water movements and change to a new equilibrium. At the sea surface and close to the continents the seasonal melting and freezing of sea ice, and river water ensure a high variability, to which the rule applies:
  - Warmer water is lighter than colder water;
  - More saline water is heavier than less saline water.
- During the winter season the ocean is always fully covered by sea ice, and as the sun does not raise above the horizon the direct influence of sunray is nil, while correspondingly during the summer season considerable.
- The Arctic Ocean is permanently supplied with new water from the Gulf Current, which enters the sea close at the surface near Spitsbergen. This current is called the West Spitsbergen current. The arriving water is relatively warm (6 to 8°C) and salty (35.1 to 35.3%) and has a mean speed

of ca. 30 cm/sec-1. The warm Atlantic water represents almost 90% of all water masses the sea receives. The other ca. 10% come via the Bering Strait or rivers. Due to the fact that the warm Atlantic water reaches usually the edge of the Arctic Ocean at Spitsbergen in open water, the cooling process starts well before entering the Polar Sea and is therefore a central aspect throughout this work.

- A further highly significant climate aspect of global dimension is the water masses the Arctic releases back to oceans. Actually, the outflow occurs only via the Fram Strait between Northeast Greenland and Spitsbergen, but together with very cold water from the Norwegian Sea basin the deep water spreads below the permanent thermocline into the three oceans, where it is slowly lost to the surface layer through weak but continuous upwelling, although it can be identified in the abyssal layers of the Pacific Ocean. However, this aspect has presumably not any traceable impact on the early arctic warming and can for the further work be neglected.

Although there is hardly a convincing reason to neglect the recent warming in the Arctic and the extent of ice melt during the summer season, it is not necessarily clear yet, whether the current discussion is based on a sound and comprehensive assessment. Climate research should not only deal with Arctic warming based on observations made during the last few decades, but at least be extremely interested in other climatic events that occurred in modern times, especially if somehow in connection with the situation in the Arctic. Why?

b) A climate revolution in the Arctic?

Beginning around 1850 the Little Ice Age ended and the climate began warming. For a long time, at least since 1650 which marked the first climatic minimum after a Medieval warm period, the Little Ice Age brought bitterly cold winters to many parts of the world, but is most thoroughly documented in the Northern Hemisphere as Europe and North America. The decreased solar activity and the increased volcanic activity are considered as causes. However, the temperature increase was remote and once again effected by the last major volcanic eruption of the Krakatoa in 1883. Up to the 1910s the warming of the world was modest.

Suddenly that changed. In the Arctic the temperatures literally exploded in winter 1918/19. The extraordinary event lasted from 1918 to 1939 is clearly demonstrated in the graph showing the 'Arctic Annual Mean Temperature Anomalies 1880 – 2004'. But this extraordinary event has a number of facets, which will be raised and discussed in due course. Meanwhile almost a full century has passed, and what do we know about this event today? Very little! Scientific literature is quite superficial concerning all three questions: Where? When? Why? For example, one can find many graphs showing the huge temperature rise in the late 1910s, but the rise was actually located in the most northern part of the North Atlantic, during winter, a location remaining sea ice free throughout the winter season, in the west of Spitsbergen.

B. A Climatic revolution at Spitsbergen and the first assessments

The enormous temperature rise at Spitsbergen in the late 1910s settled only slowly in the scientific community of those days. That may have partly be due to the fact that the Spitsbergen data recording only

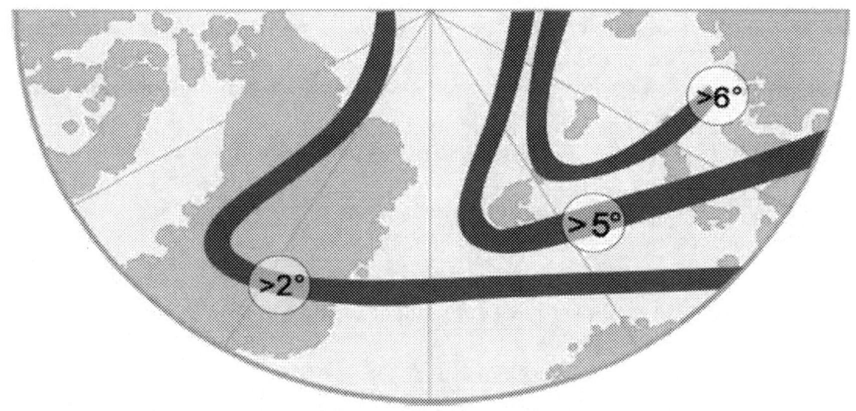

Winter temperature difference 1921-1930 minus 1911-1920
Data info source: H.H. Lamp; in: "The Arctic Ocean" (1982), Fig.7.10a.
- By approximation only - 2007AB

www.arctic-warming.com

1914

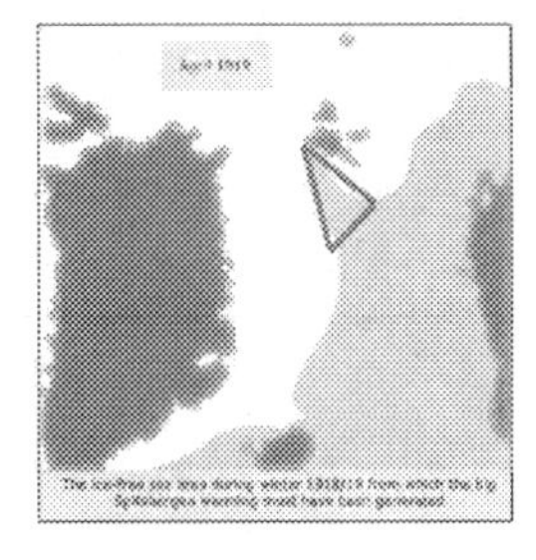

1915

1916

Sea Ice
In April 1914 /19
Did the big
warming came
from the open sea
off Spitsbergen?

1917

1918

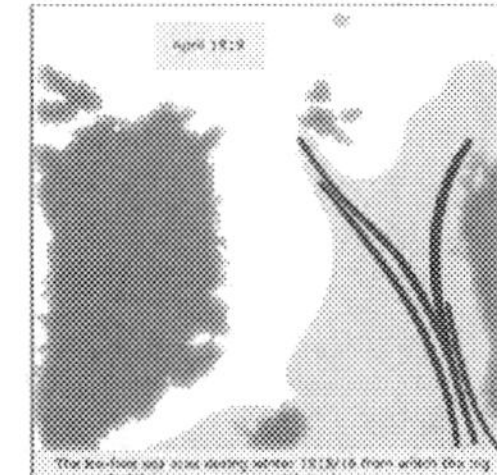

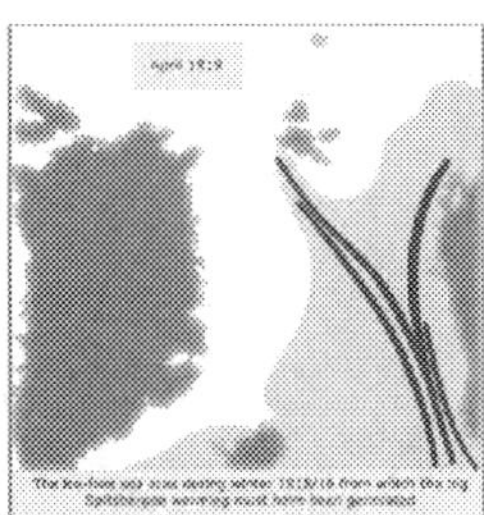

1919

started in 1912 and the interest started when the actually observed warming needed to be addressed. As much as this investigation could observe, the Spitsbergen observations have been first published by Birkeland who has already been cited with his remark: "In conclusion I would like to stress that the mean deviation results in very high figures, probably the greatest yet known on earth" (Birkeland, 1930). The discussion which followed was interesting although it was not be able to come up with convincing findings. The scientific elaborations in the 1930s will be given all deserved attention in the following chapters. But now, as a sort of warm up, there are some observations and conclusion made by O.V. Johannsson (Johannsson, 1936). Although his investigation primarily focusing on the relevance of sun-spots, some general findings are nevertheless interesting, for example:

- In 1919 the statistical means crosses zero-value; or in other words, all previous years are colder, all later years are warmer (p. 86).
- The climate had become more maritime (p. 86).
- Between 1917 and 1928 the increase during the summer season is +0.9°C per 10 years, and in winter +8.3°C, in February +11.0°C (p. 87).
- There was a colder period from 1912 to 1917 (p. 90), which, had this not occurred, would have resulted in a 1.1°C increase at the Green Harbour Station (p. 91).
- As it is known, the winters in Europe over recent decades (after 1880, even more since 1900) have become milder, the climate more maritime, the annual temperature means higher (p. 91).
- It seems that the changes are coming from the North, but this is not necessarily confirmed by temperature observations at some stations (e.g. Stockholm, Edinburgh), showing a warming from 1876-1920, but not later (p. 91).
- Temperatures in North Norway show no change between 1891-1905, but a +0.4°C change between 1921-30 (in Svalbard, 2.5°C), indicating that the increase in N-Norway is only delayed, and presumably also in Svalbard (p. 91f).

Johannsson's main conclusion is that the increased air circulation (15 % higher) as between 1896 and 1915 had gradually changed the current and ice conditions, thereby altering the borders between the Arctic gulf current climate and the true Arctic climate further north. Whether any of Johansson's findings and conclusion can withstand, a review is not so important, but the listed items show that the matter was seriously and competently discussed, and that the available material and knowledge allowed a fruitful investigation since a long time ago.

C. The early arctic warming and modern assessments

Many scientists confirm broadly the early two decade long warming period (WHEN) but fall short of identifying the exact time period and location, of which a few are here presented exemplatory:

- The warming in the 1920s and 1930s is considered to constitute the most significant regime shift experienced in the North Atlantic in the 20th century (Drinkwater, 2006).
- The huge warming of the Arctic that started in the early 1920s and lasted for almost two decades is one of the most spectacular climate events of the 20th century (Bengtsson, 2004).
- At least Polyakov (2002) get the timing right: The period from 1918 to 1922 displays exceptionally rapid winter warming not only in the circum-Arctic region northward of 62°N. (Polyakov, 2002).

- A meridional pattern was also seen in the late 1930s with anomalous winter (DJFM) SAT, at Spitsbergen (Overland, 2008).
- Average Arctic temperatures increased at almost twice the global average rate in the past 100 years. Arctic temperatures have high decadal variability, and a warm period was also observed from 1925 to 1945. (IPCC, 2007)

When it comes to explaining the causation of the warming (WHY), the matter seems rather sketchy than well founded. Here are a few examples:

- Natural variability is the most likely cause (Bengtsson, 2004);
- We theorize that the Arctic warming in the 1920s/1930s was due to natural fluctuations internal to the climate system (Johannessen, 2004).
- The low Arctic temperatures before 1920 had been caused by volcanic aerosol loading and solar radiation, but since 1920 increasing greenhouse gas concentration dominated the temperatures (Overpeck, 1997).
- The earlier warming shows large region-to-region, month-to-month, and year-to-year variability, which suggests that these composite temperature anomalies are due primarily to natural variability in weather systems (Overland, 2004).
- A combination of a global warming signal and fortuitous phasing of intrinsic climate patterns (Overland, 2008)

Further details will be provided later. At this stage it was intended to show that almost one century has passed since the first arctic warming started, and the discussion seems unable to get even the most basic facts clearly presented. A comprehensive picture of facts is paramount to consider convincing solutions. The following investigation tries to offer clues and explanations about what actually happened in the Arctic realm at the end of the 1910s and what may have caused the arctic warming at the beginning of the last century.

D. What is up for discussion?

While three issues are hotly debated worldwide: Climate Change, Global Warming, and Anthropogenic Forcing, the decisive role the oceans and seas have, has received little attention. Understanding the Arctic warming in the early last Century and the impact the oceans have played in to make the event happen, would significantly contribute to a more fruitful discussion of each of debated subjects.

The fact is that the winter temperatures made a jump of more than eight degrees Celsius at the gate of the Arctic Basin, after 1918. Nowadays, one century later, the event is still regarded as "*one of the most puzzling climate anomalies of the 20th century*" (Bengtsson, 2004). This shows that there hasn't been any convincing progress on understanding the climatic change issue! Since the Norwegian scientist B.J. Birkeland had published the temperature anomalies for Spitsbergen and concluded that these deviations could "*probably be the greatest yet known on earth*" the matter could be investigated. Although more than a ¾ century has passed the Intergovernmental Panel on Climate Change (IPCC) mentioned recently that the "*average Arctic temperatures increased at almost twice the global average rate in the past 100 years" (IPCC, 2007),* and that this represents a very significant part of the global warming issue. That is of little help if not a straight disservice to the climate debate.

1958

Th. Hesselberg & T. Werner Johannessen, Det Norske Meteorological Institute, Oslo

"The Recent Variations of the Climate and the Norwegian Arctic Sector"

in: R.C. Sutcliffe (ed): Polar Atmosphere Symposium, London 1958; p. 18f

EXTRACT from Summary:

The Norwegian Meteorologicial Service has erected five meteorological stations in the Atlantic sector of the Arctic regions, namely Spitsbergen (1912), Jan Mayen (1920), Bjørnøya (1921), Myggbukta (1922) and Hopen (1946).

- The present amelioration of the climate in the Arctic area therefore must begun before the year 1912. There are evidence that seem to indicate that it began in the 1870-years, approximately at the same time as the temperature rise began in more southerly regions.
- The rise of the temperatures in Spitstbergen is large compared with the rise in other parts of the world (about five times as great as in Norway). This fact can be explained by the position of Spitsbergen at the southern border of the inner Arctic area. This inner zone is retreating towards the north in connection with the general heating, and Spitsbergen that formerly was situated north of the parth of the cyclones surrounding the inner Arctic zones is now more frequenly visited by cyclones that bring mild air from south and south-west. Of special importance is the augmenting cyclonic winds sweep away the surface layers with very cold air that in calm weather is produced by the outgoing radiation.

EXTRACT from p. 22/23

- Of special interest are the data from Spitsbergen station, which was erected in 1912, some years before the great change in temperture conditions took place. This happened in the years 1917-1922.
- The remarkable rise in temperature is clearly seen in Fig.2 (shown to the left), that gives the departures of the mean temperatures from those of the reference period (1912-1930) at Isfjord R.. A smoothing by hand gives a rise of about 7 degrees for the winter and later on a slow increase of about 1 degree. For the other seasons the sudden rise around the year 1920 is approximately 3 degrees for the spring, 2 degrees for the summer, 3 degrees for the autumn and 4 degrees for the whole year.
- We see that the rise in the temperatures at Isfjord R. is confirmed by the data from Vardø. Also here the increase is especially rapid about the year 1920, but the rise is more modest.

Graphics: Lower figure is a cut from the Original (Fig. 2), the middle is a modified layout based on the original.

Differences between annual mean temperature
For the period 1901-1930 versus average 1859-1900

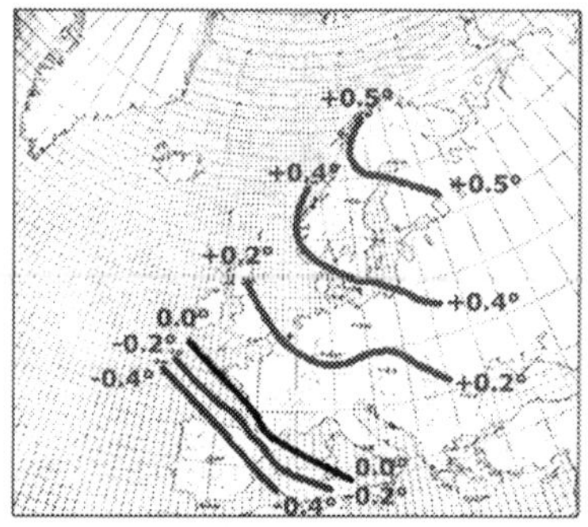

Source: Andres Angström, 1939, Geogr. Ann. XXI — 2008/www.seaclimate.com

Deviations from mean temperatures
(Dec-Feb & Jun-Aug) at Isfjord R., SPITSBERGEN

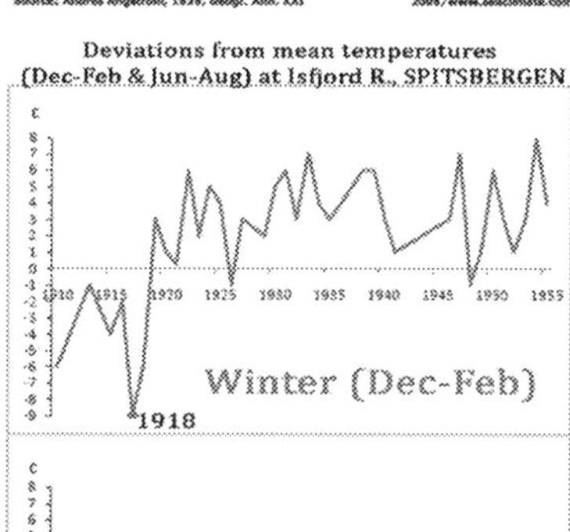

Image data based on source: Hesselberg & Johannessen, in: "Polar Atmosphere Symposium" (Oslo 1956), London 1958

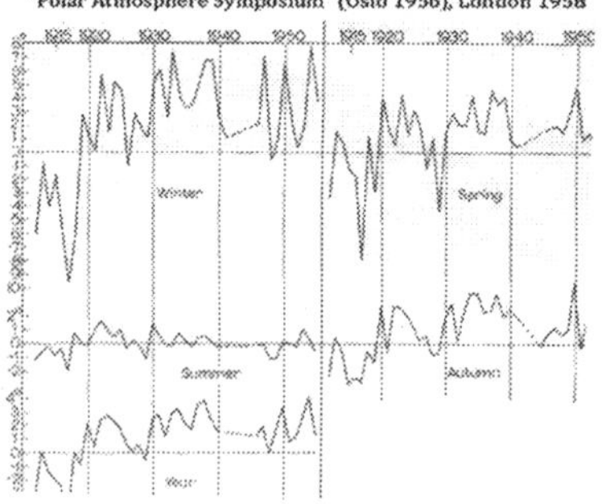

2. The Arctic in Climatology

What need to be acknowledged is that the 'hot issue' was measured in the winter of 1918/19 on the remote Norwegian island of Spitsbergen, just 1,000 kilometres away from the North Pole. Spitsbergen experienced an increase of winter temperatures of more than 8° Celsius within only a few years. When the event started, the actual figure was particularly high (see: graph), and spread out far beyond the local region over a short period of time.

But if IPCC acknowledges the arctic warming, why do we still need to go back to the beginning of the 20th century to find out the reasons of the 'Big Warming' at Spitsbergen? Isn't it enough to accept modern scientific theories, which point to "internal variability of the climate system", including the naming of three possible mechanisms, namely: anthropogenic CO_2 effects, increased solar irradiation, and reduced volcanic activity? (Bengtsson, 2004) Certainly not! The warming at Spitsbergen is one of the most outstanding climatic events since the volcanic eruption of Krakatoa, in 1883. The dramatic warming at Spitsbergen may hold key aspects for the understanding how climate ticks. The following elaboration intends to approach the matter from different angles, but on a straight line of thoughts, namely:

> **WHERE:** the warming was caused and sustained by the northern part of the Nordic Sea in the sea area of West Spitsbergen the pass way of the Spitsbergen Current.
>
> **WHEN:** The date of the commencement of warming can be established with high precision of few months, and which was definately in place by January 1919.
>
> **WHY:** the sudden and significant temperature deviation around the winter of 1918/19 was with considerable probability caused, at least partly, by a devastating naval war which took place around the Great Britain, between 1914 and 1918.

a) The objective of this investigation

The objective of investigating the Arctic Warming from 1918 to 1940 is to demonstrate that the oceans and seas control and determine the global climate. Insofar it is a part of a wider analysis[20] concerning the four major climatic diversions during the 20th Century, namely:

A. The Arctic Warming from 1918 to 1940;
B. The extreme cold war winters in Northern Europe, 1939-40, 1940-41, and 1941-1942;
C. The global cooling from ca. 1942 to ca 1970;
D. The resume of the pre WWII warming, or was this warming totally separated from previous trends?

What makes the early Arctic warming interesting and challenging is the necessity to establish at least two facts first, namely: Where and When, before considering any probable causation: WHY.

Imagine that a 'revolution' took place 90 years ago, but "many" only talk about future climate changes, and some even claim that they can predict the climate in 10, 20, or even 100 years ahead, having no idea or explanation what had happened not very long ago. That seems hardly acceptable, when there is a case at

[20] See e.g.: www.seaclimate.com; www.warchangesclimate.com; and related book publications. Essays from 1992 to 1997: www.oceanclimate.de.

2008

Joe D'Aleo claims:

Warming in the arctic is likewise shown to be cyclical in nature.

Discussed at: http://www.arctic-warming.com/; 17 May 2008

Joe D'Aleo claims:
Warming in the arctic is likewise shown to be cyclical in nature.
- An assertion that can be challenged –

In a recent article „Multidecadal Ocean Cycles and Greenland and the Arctic" by Joe D'Aleo (http://www.intellicast.com/Community/Content.aspx?a=128) on the 12th of May 2008, the author says:

"This week we will talk about temperatures and ice in Greenland and the Arctic, topics sure to dominate the news this summer. Already recent media stories have some scientists predicting another big melt this summer. We will show how that is not at all unprecedented (happens predictably every 60 years or so) and is in fact entirely natural", and

"We will show how that is not at all unprecedented (happens predictably every 60 years or so) and is in fact entirely natural."

The readable and interesting paper should not go unchallenged. Joe D'Aleo concludes i.a. that: *"The warm mode of the Atlantic Multidecadal Oscillation (AMO) also produces general warmth across much of the Northern Hemisphere including Greenland and the Arctic."*

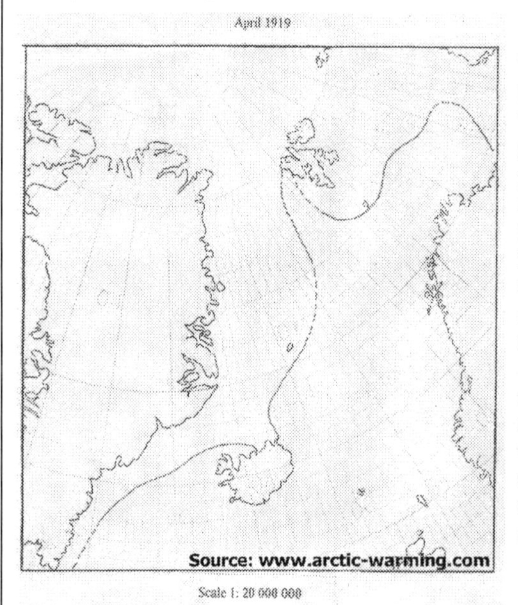

The fact is that the early arctic warming was anything else but not cyclical. It was an "explosion" and not gradual shift. The extreme rise of temperatures was initially confined to Spitsbergen, and only commenced in Greenland at least one year later. The sea ice cover off Greenland's coast had been reaching Iceland in April 1919 for the first time since 1911 (see b/w graphic). The research of this site and the PACON 2007 Conference Paper show that the early warming has nothing to do with Atlantic Oscillation, but had been entirely related to the impact of the arm of the Golf Current that passed Spitsbergen prior to entering the Arctic Ocean. Actually, the extensive sea ice cover in the North Atlantic until the month of April, prevented significantly to produce more warm air, which could have generated the extreme winter temperature rise at Spitsbergen, the remote archipelagos almost surrounded by ice up to April 1919, except a small open sea area formed like a tongue extending almost to the Arctic Ocean.

simple. The annual arctic average temperature swelled 2.5° Celsius between 1918 and 1939, the winter temperatures were much higher. But neither the general attribution of this event to the Arctic region, nor the reference to a time period over one or two decades are sufficient to get a clear picture of this extraordinary climatic event. In the next sections we will outline the principal parameters on which the subsequent discussion will take place.

b) WHERE?

The boundary of the Arctic is generally considered to be north of the Arctic Circle, which runs along the latitude of 66° 33' North, comprising an area of about 1.5 times the size of the United States. What happened in one corner of this huge area did not necessarily occur in another corner as well.

Speaking only generally about arctic warming may disguise important information. The 'climatic revolution' is precisely such case. The sudden warming at a very confined location, the remote archipelagos of Spitsbergen at the latitude of 77-80° North, or about 1000 km north of the Nord Cape of Norway is a key factor for analyzing the Arctic Warming in the late 1910s.

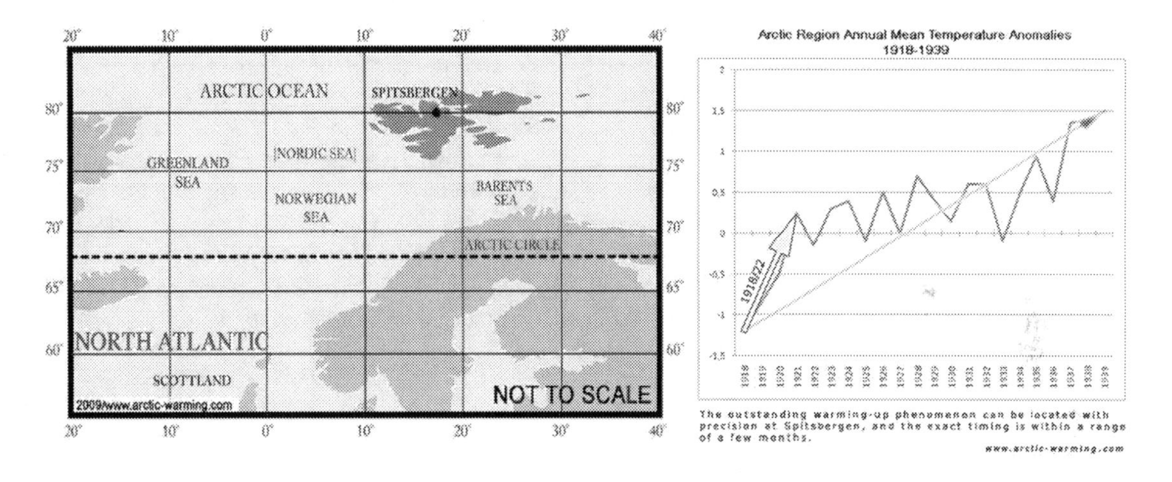

The outstanding warming-up phenomenon can be located with precision at Spitsbergen, and the exact timing is within a range of a few months.

www.arctic-warming.com

The overriding aspect of the location is the sea; the sea around Spitsbergen, the sea between particularly the Norwegian-, the Greenland-, and the Barents Sea (Nordic Sea). The Norwegian Sea is a huge, 3000 metres deep basin. This huge water masses stores a great amount of energy, which can transfer warmth into the atmosphere for a long time. In contrast the Barents Sea, in the southeast of Spitsbergen has an average depth of just around 230 metres. In- and outflow are so high that the whole water body is completely renewed in less than 5 years. However, both sea areas are strongly influenced by the water masses coming from the South. The most important element is a separate branch of the North Atlantic Gulf Current, which brings very warm and very salty water into the Norwegian Sea and into the Spitsbergen region. Water temperature and degree of saltiness play a decisive role in the internal dynamics of the sea body.

And what might be the role of the huge basin of the Arctic Ocean, 3000 metres depth and a size of about 15 million square kilometres? The difference towards the other seas mentioned is tremendous. The Arctic Ocean used to be widely ice covered in the first half of the 20th Century, the other seas only partly on a seasonal basis. Only between the open sea and the atmosphere an intensive heat transfer is permanently taking place. Compact sea ice reduces this transfer about 90% and more, broken or floating ice may change the proportion marginally. In this respect an ice covered Arctic Ocean has not an oceanic but 'continental' impact on the climate.

c) WHEN?

This investigation is fully aware that the sudden rise of arctic temperatures could be a mere coincidence of circumstances, but is unwilling to accept such an approach without challenging it. An explanation is necessary because this event didn't come from nowhere; it must have been caused by a physical force and dynamics, which resulted in sudden and unexpected temperature rises. Fortunately the collection of temperature data had already started in the early 20th Century. Already in the 1930s and 1940s the phenomenon was analyzed, albeit on a long-term statistical basis only.

Although it is long known that "*in 1919, the statistical means crosses zero value; or in other words, all previous years are colder, and all later years are warmer (*Johannsson 1936) one need to try to analyze the timing much more precisely.

This will be done thoroughly, whereby the suddenness and the value of increase at the time of commencement of the event and during the following time period will be given particular attention. In so far it surprises that Polyakov (2003) states that *"the period between 1918 and 1922 displays exceptionally rapid winter warming",* but did not elaborate this aspect any further. We will demonstrate that warming started within a few months' period in 1918, latest in January/February 1919. The precise timing is, as the location, a decisive aspect to consider the causation of the event.

In this respect it is to note that the investigation rests on observed wintertime temperatures. This is a decisive tool. It excludes very clearly any direct influence of the sun and carbon dioxide (CO2), due to a long polar night. From October until February, sun radiation is virtually inexistent at Spitsbergen, and its direct influence on temperatures is practically zero. The winter season also diminishes any claim that CO2 could have played any significant role.

d) WHY?

Once the location and time of the commencement of the arctic warming have been determined with high precision, and that will be done, the next question inevitably arises: WHY? After all, the matter concerns a 'climatic revolution'. No earthquake shook the earth crust, no volcano of significance had erupted, no meteorite had hit the Earth, neither had any tsunami been observed. We will certainly not exclude the option that nature did it on "its own".

But this conclusion can and should only been drawn, if a serious contender for having caused the arctic warming since winter 1918-19 has evidently been excluded, namely the naval war around Great Britain and at other Northern European sea areas from 1914 to 1918. Not only was World War I presumably the biggest interference to the natural commons since Krakatoa, but most of the sea water which had been subject to naval war activities did not remain in their place but travelled with sea current system into the

Spitsbergen region were suddenly the temperatures exploded in winter 1918-19. The thesis will be elaborated in depth in Chapter 8: "What caused the Arctic Warming?"

E. Expected results

Current conclusions, as outlined above, will be challenged at least on the ground that:

The "Big Warming" event from Spitsbergen proves that the climate change was not determined by the atmosphere; but first of all, if not alone by the sea.

If climate were defined as 'the continuation of the oceans by other means'[21], the arctic warming would – with high certainty – have been fully understood and explained since long, and thus could have provided a big service to the general climatic change debate.

[21] A number of papers by Arnd Bernaerts since 1992 suggest to use this term; see:www.oceanclimate.de , www.whatisclimate.com

1937	**Scherhag, R., (1937) ‚Die Erwärmung der Arktis',**
	in: Cons. Intern. Expl. Mer. Rap. Proc.- Verb., Copenhagen, Vol. 12, p. 263-276.

SUMMARY in ENLISH
The greater mildness of winters observable in the temperate zone during the last hundred years, accompanied by an increase in atmospheric circulation, has, during the last fifteen years, led to an extraordinary rise in temperature in the arctic regions, which in its turn has been accompanied by a corresponding retreat of the ice and a higher temperature in the sea. Just as, in the region from which the Gulf Stream springs, measurements of water temperature indicate a rise of about 0.5°C in the last ten years, there has also been found a like increase in surface temperature in the English Channel. Fifty-years series of temperatures from off-lying Norwegian coastal stations clearly manifest a similar warming, and it is therefore indubitable that the transport of the warm water of the Gulf Stream from Florida to its entry into the arctic has increased to a noteworthy extent. In this respect it appears to be a question of a secular period in the variation of atmospheric circulation of some 225 years duration, which seems at this juncture to have attained its maximum.

Tabelle 3.
Mittlere Abweichung der Eisgrenze (in km.) im Ostspitzbergenmeer
(30° bis 50° Ostlänge) im Spätsommer.

Jahr	Abweichung	Jahr	Abweichung	Jahr	Abweichung
1898	-140	1911	0	1923	-210
1899	-100	1912	+110	1924	-100
1900	-50	1913	+160	1925	-170
1901	-60	1914	+120	1926	-50
1902	+40	1915	+30	1927	-10
1903	+80	1916	+320	1928	-70
1904	+10	1917	+140	1929	-10
1905	-110	1918	+100	1930	-290
1906	-150	1919	-30	1931	-270
1907	-240	1920	-140	1932	-100
1908	-230	1921	-120	1933	-280
1909	+70	1922	-270	1934	-180
1910	+80				

Late summer: plus = more ice-cover; minus = less ice-cover
EAST-SPITSBERG-SEA; 30° to 50° East

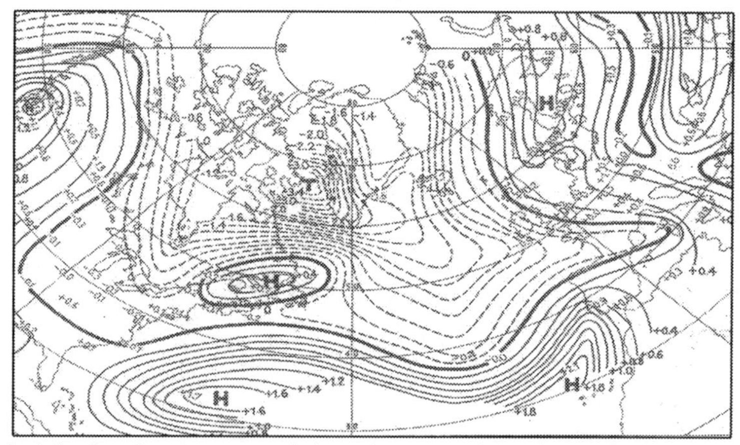

Deviation from long-term mean of winter air pressure during the decade 1921-1930.
Scherhag, R. (1936/Sept.); "Die Zunahme der atmosphärischen Zirkulation in den letzten 25 Jahren"; Annalen der Hydrographie und Maritimen Meteorologie, p. 397-407, Fig 7 (here reduced and streched)

Chapter 3
Spitsbergen temperature rocketing

A. Factual Aspects

a) A rocket rise

The rocket rise of the winter temperature at Spitsbergen is well illustrated in a graphic that the Norwegian scientists published more than 50 years ago (see p.3). In the oral presentation at the 'Polar Atmosphere Symposium" held in Oslo from 2-8 July 1956 the authors Hesselberg and Johannessen[22] explained:

> *"Of special interest are the data from Spitsbergen where the series of observations go back to 1912. During the first years the observations shows no conspicuous climatic change, but then comes a rapid rise of the temperature in the year 1917 to 1922. The increase of the mean temperatures in this period was about 7 degrees Celsius in the winter, 3 degrees in the spring, 3 degrees in the summer, 3 degrees in the autumn and 4 degrees for the whole year. After the year 1922 the temperature continued to rise until the war broke off the series, but the rise was much slower."*
> *"The rise of the Temperature in Spitsbergen is large compared with the rise in other parts of the world (about five times as great as in Norway). This fact can be explained by the position of Spitsbergen at the southern border of the inner Arctic area".* (Hesselberg, 1958)

Not less impressive is a further graph showing the temperature developments in Norway from Spitsbergen to Oslo during the years 1871 –1938, (Manley, 1944)[23]. The image indicated the changes in the ten-yearly winter mean temperature in Tromso, Röros, Bergen, Oslo, and Spitsbergen. It is worth to observe the time of commencement of the rise at different locations, showing that the turning point was later in the South of Norway as in the North: Spitsbergen before 1920; Tromso in 1920, Röros ca. 1921-25; and in Bergen and Oslo, 1924-25. While Manley (1944) points at the fact that "Temperature in Norway, especially in the North, has certainly risen far more in recent years than at any other time in the last two centuries", Johannsson (1936) confirms that the increasing temperatures have been coming "from the North". As the 'rise' sustained for two decades, only the seas, by a substantial shift of the seawater bodies around Spitsbergen and the Northern Seas, could have generated such long-term climatic changes. This section attempts to establish that a colossal temperature rise occurred in the Spitsbergen region from summer 1918 to winter 1918/19.

b) Scope of data and other investigation sources

One aspect is official: During the last century the increase of temperatures in the Arctic was two times higher than the global average (IPCC, 2007), which is an interesting aspect but explains little. A detailed

---

[22] See also: Special Page at Chapter 2, and SP (Overland, 2008) this Chapter.
[23] Chapter 5, section e) Europe

analysis needs more elaboration. The immediate problem is that there is almost a complete lack of sea water temperatures, not only from lower level, but neither from the sea surface (SST). The pillars for researching the climatic developments are surface air temperatures (SAT), and also they are rare during those days.

But at least in one case the Arctic research was in luck. At Spitsbergen, the first permanent temperature data series began in 1912, right in time for recording the presumably highest temperature rise ever observed (Birkeland, 1930). Few further places within the Polar Circle provide data as well, and are now available at the website of NASA/GISS[24], e.g. since 1880 in West Greenland, Upernavik and Jakobshaven, and Grimsey in Iceland. In most places in the Nordic Sea area, e.g. East Greenland, Jan Mayen, and Bear Island, weather records were taken only since 1920 or later. Actually, for the first quarter of the last century, solid data concerning the polar region is limited and has to rely on few expedition records and interpretation of secondary observations.

Another investigation source is, probably, the change of the ocean ecosystem, illustrated by a graph showing the great increase of the cod fishery of the West-Greenlanders since ca. 1926 (at Chapter 5), due to the higher water temperatures (Carruthers, 1941). This phenomenon in the northern North Atlantic has been subject for a number of papers since long (e.g. Lee, 1955).

More recently Drinkwater concluded from such changes in fish populations occurring during the 1920s and 1930s can be linked to a general warming of the oceans, not only due to a large rise in air temperatures alone, but to an apparent change in ocean circulation that brought more warm water northwards (Drinkwater, 2006). This is an interesting and supportive material, but not the explanation needed for the way it happened. This investigation will leave the changes in the ecosystem aside and concentrate on the available temperature data.

c) Which data used are promising

For climate research the location of Spitsbergen is a blessing, and not only in one respect, namely:

- Located in the middle of three huge water bodies in volume and size:
    - Norwegian/Greenland Sea;
    - The Arctic Ocean; and
    - The Barents Sea, with a modest volume (mean depth ca. 280m) but considerable size.
- Located at the edge of sea ice, were regardless of the time of a season at least a tiny space of the sea remains ice free, which ensures a maritime induced climatology, while a space covered with sea ice induces continental climatology.
- Located were the sun does not rise above the horizon for the whole winter period, at Spitsbergen from the 26th of October to the 16th of February (Birkeland, 1930).

The last point would make climate research much easier, because from two ruling elements of climate, the sun and water, the sun can be neglected of having a direct impact over a couple of months. The citing of

[24] NASA, Goddard Institute for Spaces Studies/NY; http://data.giss.nasa.gov/gistemp/station_data/

2004

Johannessen, Ola M., et al. (2004);

"Arctic climate change – Observed and modeled temperature and sea ice variability",

Tellus 56A , p. 328 –341, Corr. 559-560.

A selection of aspects from the paper Abstract read as follows:

1. Changes apparently in the arctic climate system in recent years require evaluation in a century-scale perspective in order to assess the Arctic's response to increasing anthropogenic greenhouse-gas forcing.
2. We show that two pronounced 20th-centrury warming events, both amplified in the Arctic, were linked to sea-ice variability.
3. SAT observations and model simulations indicate that the nature of the arctic warming in the last two decades is distinct from the early 20th-centrury warm period.
4. It is suggested strongly that the earlier warming was natural internal climate-system variability.

WINTER HALF-YEAR SURFACE TEMPERATURES TREND, 1920-1939

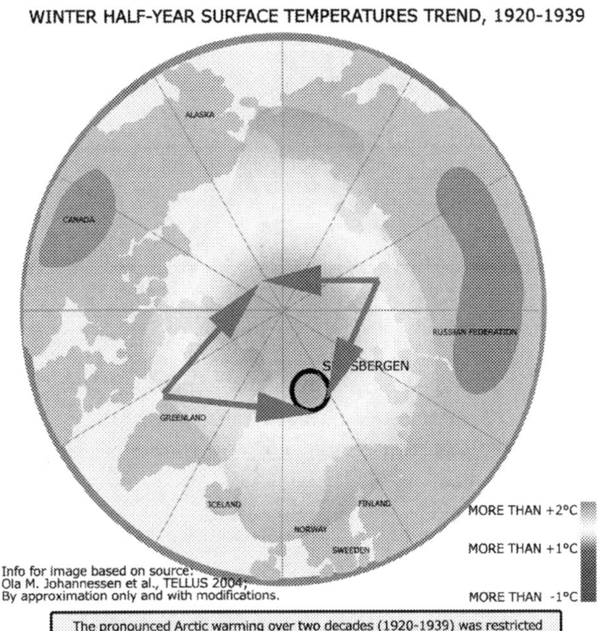

The pronounced Arctic warming over two decades (1920-1939) was restricted to an area not larger than ca. 1/3 of the whole Arctic Ocean. 2007AB

Comment: Figure 2 of the paper includes an image of seasonal SAT trends north of 30°N. The general indication for a 6 months winter season for the two decades 1920-1939 (which are in great conformity with the R. Scherhag data from 1936, and H.H. Lamp from 1982) the following graphic has been prepared. These and other graphics show, that the intensive early warming was not throughout the Arctic, but only in the North Atlantic sector. One of the co-authors, V.F. Zakharov noted this already 1997, as mentioned in Chapter 1, by saying: "*Why are the maximum climate fluctuations confined to the Atlantic sector of the Arctic?* (Zakharov, 1997). Neither he, nor any of his 11 (et al.) colleagues pay any attention to this aspect. Although they assume sea-ice variability as applicable to the early warming, they do not even realize that the early warming commenced in 1918/19 despite the fact that the winter sea ice was not reduced (see: April sea-ice in Chapter 2), and that they should have at least acknowledged the suddenness of the temperature rise since winter 1918/19. But as their oldest reference material dates from 1982 (Kelly, 1982), they ignored all research material published over 50 years since 1930.

the following paragraph, taken from a press release of the Max Planck Institute for Solar System Research[25], may illustrate why this is a serious aspect:

> *"The influence of the Sun on the Earth is seen increasingly as one cause of the observed global warming since 1900, along with the emission of the greenhouse gas, carbon dioxide, from the combustion of coal, gas, and oil. "Just how large this role is, must still be investigated, since, according to our latest knowledge on the variations of the solar magnetic field, the significant increase in the Earth's temperature since 1980 is indeed to be ascribed to the greenhouse effect caused by carbon dioxide," says Prof. Sami K. Solanki, solar physicist and director at the Max Planck Institute for Solar System Research."*

To find out at what time exactly the climatic changes of the 1920s started, the following discussion considers the core winter months of December to February, if not stated otherwise. Tracing the sources of 'climate making' is much easier if the sun is not involved. Without the sun heat from the oceans is the sole sustainer of the weather mechanism in wintertime at high latitude.

B. The heating up of Spitsbergen

a) A sudden shift

As mentioned earlier, the information given for Spitsbergen (Svalbard) by Birkeland in 1930 was already a quite sufficient indication of the temperature shift. The change came with suddenness. On the basis of half a dozen years the jump before and after winter 1918/19 is about 8°C. Comparing only January/February of 1917 and 1918, with January/February of 1919 and 1920 the temperature jump is almost plus 10°C.

During the winter of 1918/19 the temperatures varied much. There were long periods in November and December 1918 with close to zero degrees (approx. 26 days less than 5°C), with 4 days above zero in November and 7 days in December. In January 1919, on 14 days the temperatures did not reach –5°C, five days were frost-free.

With average monthly temperatures of –7.5°C and +8.0°C, respectively, above 15-year means the sea must have transferred a lot of heat to the air. However, during February – April 1919, the temperatures were well below the average with a large ice cover far out into the sea. But that did not affect the significant warming that started a few weeks earlier.

One further point needs to be observed. Actually, the 'warming-up' process must have started some months before winter 1918/19. The annual deviation for 1918, i.e. "+0.1", indicates the end of a cooling trend since 1915, during the previous winter 1917/18, sometime in spring or summer 1918. There exists even a report that during the summer 1918 the water in the Fjords of Spitsbergen west coast had been very warm, 7-8°C (Weickmann, 1942).

---

[25] Max Planck Institute for Solar System Research (2004), "How Strongly Does the Sun Influence the Global Climate?" Press Release 8/2004, the 2nd of August 2004

3. Spitsbergen temperature rocketing

Spitsbergen

Deviation from mean

Annual

&

January – February

1912 - 1926

Year	Annual deviation	January Deviation	February deviation	Sum of Jan-Feb
1912	-3.1	-8.4	-7.3	-15.7
1913	+0,2	+0.3	-1.7	-1.4
1914	-1,3	-5,7	-4.9	-10.6
1915	-2.0	+1.8	-0.5	+1.3
1916	-2,5	-8.6	+2.1	-5.5
1917	-5.0	-7.4	-10.3	-17.7
1918	+0.1	-10.1	-0.4	-10.5
Mean deviation per winter months Jan., Feb.: - 4.3				
1919	-0.8	+8.6	-4.7	+3.9
1920	+2.3	+3.8	+1.4	+5.2
1921	+0.6	-0.8	+0.1	-0.7
1922	+2.5	+10.5	+6.9	+17.4
1923	+2.9	+3.3	+4.8	+8.1
1924	+2.5	+5.7	+8.1	+13.8
1925	+1.9	+4.3	+6.3	+10.6
1926	+0.8	+2.2	+0.5	+2.7
Mean deviation per winter months Jan., Feb: +3.8				

Average January mean temperatures

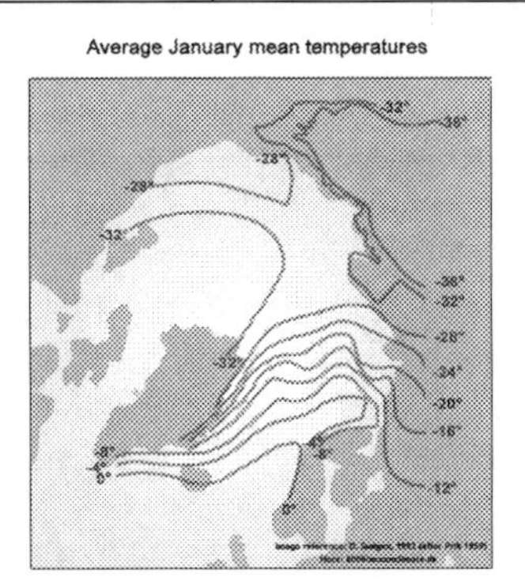

Average July mean temperatures

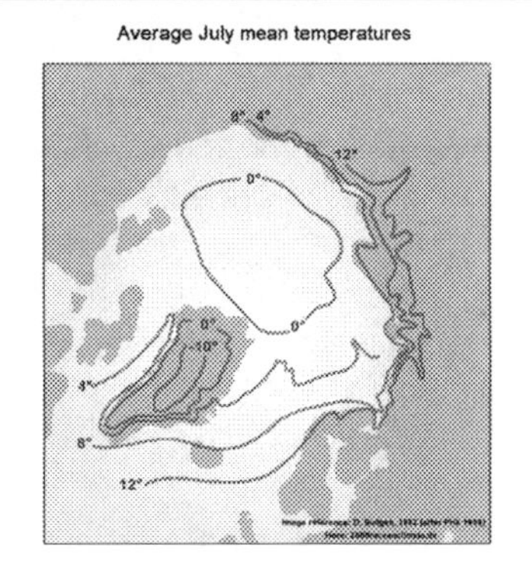

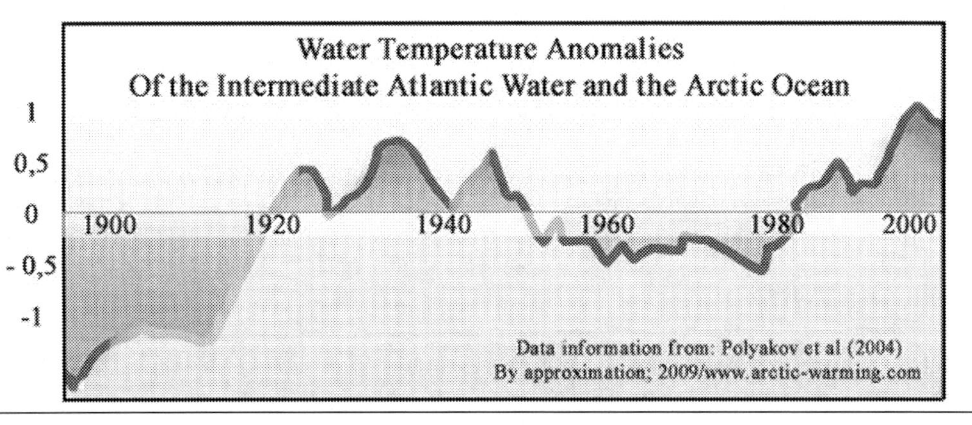

b) Spitsbergen in comments between 1930 and 1982

The presented comments are an arbitrary selection and do not mention reference already made earlier with one or two exceptions. Further details will be given in the next chapter. The listing will show that the interest in the Spitsbergen event diminished over the time instead of receiving more. Nevertheless, it can be shown that the previous generation of researchers in earth science, might have been closer to understand the Spitsbergen event as modern science, but had been stopped by the Second World War.

Birkeland (1930) The mean deviation of the Green Harbour' Spitsbergen station data results in very high figures, probably the greatest yet known on earth.

Kunz (1933) Kunz dates the temperature shift at Spitsbergen to 1918, based on the winter ice conditions around Spitsbergen, noting that after the very ice-rich years of 1915-17 the subsequent years since 1918 had most been ice-poor.

Schokalsky (1936) The discovery concerning the warming of the Polar Sea, which dates from 1921, was also observed by the 1928 *Marion* Expedition in the Baffin Bay as well as in the Barents Sea.

Scherhag (1936/Sept.) During the decade 1912 to 1930 the Northern Hemisphere shows an increase in winter temperatures, which exceeds 1° north of 60°N, at the West coast of Greenland 2-3° and even more at Spitsbergen. The areas with decreasing temperatures fall aback and are restricted to middle Asia, the western part of the Mediterranean, the Atlantic Ocean between the Azores and Bermuda. The contribution to the warming in the temperate zones but particularly in the arctic region is sheer perplexing. The general situation indicates that the changes are due to an entire change in the circulation.

Scherhag (1937) All stations north of latitude 55° North indicate a warming, which increase towards the pole, and reaching a maximum along the West Coast of Greenland with 2 degrees Celsius. However the warming is even higher at Spitsbergen. Such stipulation of temperature change as we observe at Spitsbergen, needs to be accounted as the largest climatic changes!

Brooks (1938) At Spitsbergen at least, the rise occurred in two stages, the winters of 1922-23 to 1924-25 being warm, those of 1925-26 to 1929-30 somewhat cooler, and those of 1930-31 onwards warmer than the first group.

Scherhag (1939/Feb) The temperature increase at Spitsbergen, which emerged for the first time in winter 1918/19, brought an increase of 5° over the period from 1912 to 1920. The warming got a phenomenal increase during the 1930s of 9°. The culmination of this development is not yet foreseeable: the winter 1936/37 was warmer than all previous records , and the winter 1937/38 broke this records as well, and was in average by 16° warmer than the winter 1916/17. There can be no doubt any longer that the temperature increase in the polar region represents the largest climatic change since regular meteorological observations are recorded.

Scherhag (1939/Juni) The water temperatures in West Greenland have been remarkably low in the last two years (1937 and 1938) together with colder winters. On the contrary, nothing comparable could be observed at Spitsbergen, where the mean temperatures of the last winter (November to March) superseded with a positive deviation of +8.5°C all pervious years.

3. Spitsbergen temperature rocketing

Carruthers (1941) In August, 1931, H. Mosby in the "Quest" observed much higher salinities in the Atlantic water in the polar sea north-east of Spitsbergen than had been observed by earlier expeditions.

Manley (1941) The effect was indeed remarkable; the salty Atlantic water penetrated further into the Arctic to such a degree that, for example, the average length of the coal shipping season at Spitsbergen almost doubled in length, from 95 days during 1909-12 to 175 days during 1930-38.

Henning (1949) Before 1917 the duration of shipping to Spitsbergen had averaged 94 days but since 1918 –1939 it has become 157 days. The warming moved the vegetation in Scandinavia some 100 km further north.

Kirch (1966) The world-wide rise in temperature, which began last century and has reached its climax in the thirties of this century, has been especially well-marked in the Arctic region. In the summer months, the warming was generally in the order of magnitude of 1 degree Celsius, whereas it reached 7.7 degrees Celsius in the 10 years mean over Spitsbergen in the winter months; at Spitsbergen was the annual mean of warming about 3.7 degrees Celsius. Most stations showed somewhat lower values, the main rise in temperature, however, always took place in the winter months.

Lamb 1977 The strongest (and therefore most easily established) effects on temperature are – apparently as in other, earlier climatic fluctuations – observed in the highest northern latitudes

Lamb (1982) The change of prevailing temperatures seems to be the greatest in the regions affected by changes in the balance between the warm northbound Atlantic water and the cold polar current at the ocean surface in the Norwegian – Barents Sea – east of Greenland region. Lamp provides a graphic account of the winter temperature deviation in the decade 1921-30 (minus winter 1911-20) and with the centre east of Spitsbergen (+6°C). Lamb indicates that this region, together with the Norwegian Sea, seems to be the most sensitive to climatic variations.

Jones (1982) The warming between 1881 and 1940 and the subsequent cooling to the mid or late 1960s are readily discernable.

c) How does modern science talk about Spitsbergen event?

The Russian scientist J. Schokalsky said to the Royal Scottish Geographical Society that „ it is necessary to know more about the thermal condition of the branch of the Atlantic Current which passes round Spitsbergen". That was not last year, but well before World War II in 1935 (Schokalsky, 1936). It seems that papers published in the first decade of the new millennium are silent in this respect, and mention the extraordinary Spitsbergen event not at all, or are extreme superficial on the early warming. What is even more surprising that virtually not any efforts have been made to use the findings of the pre-WWII generation, to analyse their conclusions, or at least to make references to their publications. Their reference papers are usually not older than 10 years. Prominent names in science in the 1930s, e.g. Brooks, Helland-Hansen, and Scherhag acknowledge? Why is this a complete negative return? Do they regard themselves much wiser as their predecessor? How little specific attention is given to the location Spitsbergen shall be illustrated by few examples from well known experts.

1938	Brooks, C.E.P., **"The Warming Arctic"**, The Meteorological Magazine, 1938, p.29-32.

EXTRACTS FROM PAPER

In recent years attention is being directed more and more towards a problem which may possibly prove of great significance in human affairs, the rise of temperatures in the northern hemisphere, and especially in the arctic regions.

- The Spitsbergen branch of the North Atlantic Current has greatly increased in strength and the surface layer of cold water in the Arctic Ocean has decreased from 200 to 100 metres thickness.
- Attributing the recent period of warm winters to an increase in the strength of the atmospheric circulation only pushes the problem one stage further back, for we should still have to account for the changein circulation.
- Moreover, it is almost equally plausible to regard the change of circulation as a result of the warming of the Arctic, for open ice conditions in the Arctic Ocean are favorable to the formation of depression.
- More probably the increased circulation is both cause and effect of the warmed Arctic

DEVIATION OF TEMPERATURES FROM NORMAL
JANUARY 1938

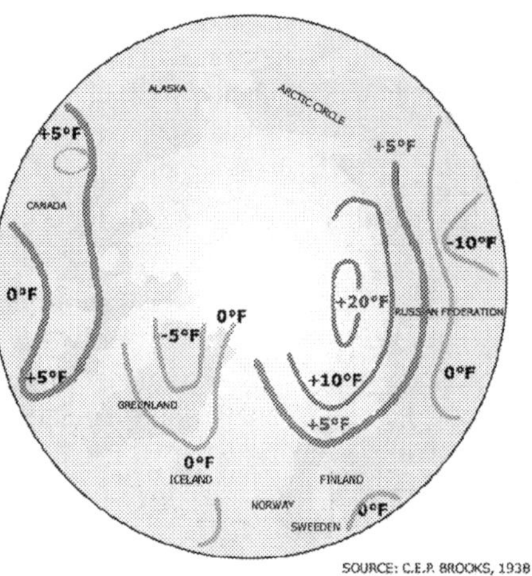

SOURCE: C.E.P. BROOKS, 1938
2009/www.oceanclimate.de

COMMENT: C.E.P. Brooks was very close to the solution of the causation of the early Arctic warming. Remarkable is that he challenged already R. Scherhag's rational that the circulation had initiated the warming, which is still in use today. Brooks seem to have overseen the suddenness by which the warming started, because this would have made clear to him that atmospheric warming followed the ocean warming, and during the winter season it could only be provided by a sudden shift in the warm Spitsbergen Current. Brooks far-sightedness was lauded by J.N. Carruthers by writing in 1941:

Eight years ago, the very wide subject: "oceanography and meteorology" was treated expertly and in considerable detail in a 60-page paper which confer a real boon on the practitioners of both our sciences. The writer was the American meteorologist C.F. Brooks, who has had wide dealings with the sea and who made very extensive investigations on ocean temperatures among other things. In one section of his valuable paper (FN), entitled "Surface oceanography fundamental to world meteorology," C.F. Brooks treats the following subjects:

__*The ocean as regulator of the world weather.*

__*The ocean and the planetary wind belts.*

__*Seasonal abnormities in centres of action.*

__*Ocean temperatures in seasonal weather forecasting.*(Carruthers, 1941) [1];

[1] C.F. Brooks "Oceanography and Meteorology", Chapter 14 (457-519) of Physics on the Earth-V. "Oceanography" Bull. Nat. Coun., Wash., No.85, June 1932.

3. Spitsbergen temperature rocketing

Polyakov (2003) a) The Arctic SAT (sea-air temperature) shows two maxima: in the 1930s –1940s and in recent decades. b) The warming in the 1920s-30s was rapid in spring and autumn and very rapid in winter, and much weaker in summer. c) The period from 1918 to 1922 displays exceptionally rapid winter warming.

Polyakov (2004) The warm and salty Atlantic water (AW) plays a special role in the thermal balance of the Arctic Ocean.

Kelly (1982) During the final years of the 1910s, warming began in the Barents Sea and Kara Sea regions. The 1920's was a transitional decade with strong warming affecting most regions. The Barents and Kara Seas had warmed by ~2°C (annual data) by the mid-1920s.

Polyakov (2004) In contrast to the warming period of the 1990s, the 1930s warm period in the Arctic did not coincide with a positive phase of the NOA (North Atlantic Oscillation).

Johannessen (2004) Two characteristic warming events stand out, the first from the mid-1920s to about 1940 and the second starting about 1980 and is still ongoing.

Bengtsson (2004) The huge warming of the Arctic that started in the early 1920s and lasted for almost two decades is one of the most spectacular climate events of the 20th century.

Drinkwater (2006) During the 1920s and 1930s, there was a dramatic warming of the northern North Atlantic Ocean, that was considered to represent the most significant regime shift experienced in the North Atlantic in the 20th century. Drinkwater makes several references to pre WWII papers, however, his subject is the ecosystem and fishery.

Overland (2005) In the early period, roughly 1920 –1927, the positive phase of the Atlantic Oscillation (AO), or more locally the North Atlantic Oscillation, had a contribution on the North Atlantic seesaw with warm temperature anomalies in Europe and cold anomalies in west Greenland.

Brönnimann (2008) Given the similarity of response in Arctic temperatures during the early and late 20th century warming, the question remains: To what extent is Arctic temperature controlled by global warming, by the regional atmospheric circulation, or by lower frequency oceanic processes?

IPY 2007-2008 Activities[26]: During last decades in the Euro-Arctic Region there is observed a stable tendency towards warming that enables us to assume that this is not a short-time deviation of the climatic system from the equilibrium but long-lasted changes.

… Spitsbergen is a wonderful science platform for studying the overall spectrum of reactions of Polar Regions nature on the climate variations both of natural and anthropogenic origin.

[26] International Polar Year (IPY) 2007-2008; Spitsbergen Climate System Current Status – SCSCS. (Abstract from Summery of Activities); http://classic.ipy.org/development/eoi/proposal-details.php?id=357

3. Spitsbergen temperature rocketing

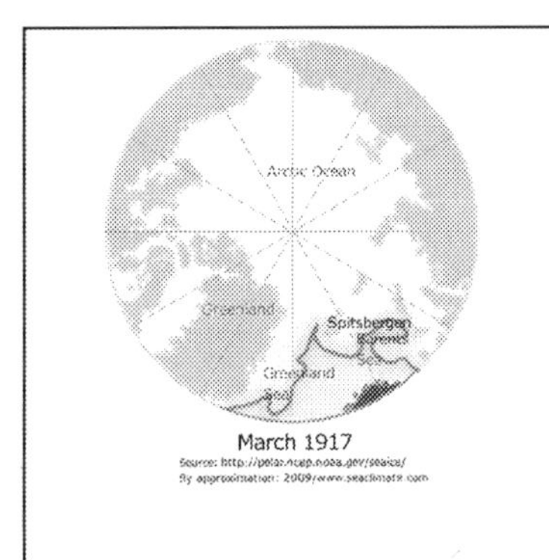

March 1917
Source: http://polar.ncep.noaa.gov/seaice/
By approximation: 2009/www.seaclimate.com

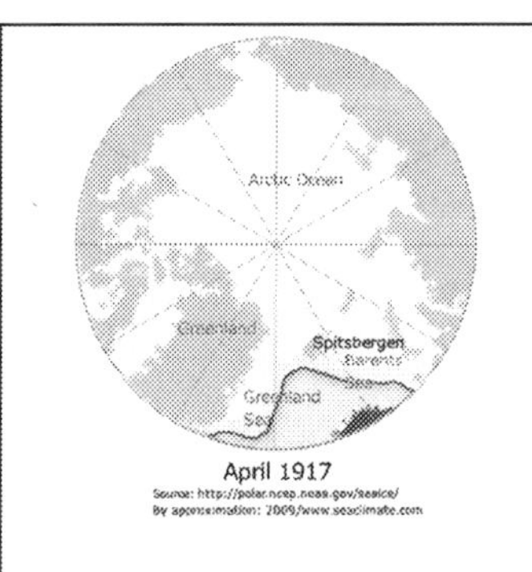

April 1917
Source: http://polar.ncep.noaa.gov/seaice/
By approximation: 2009/www.seaclimate.com

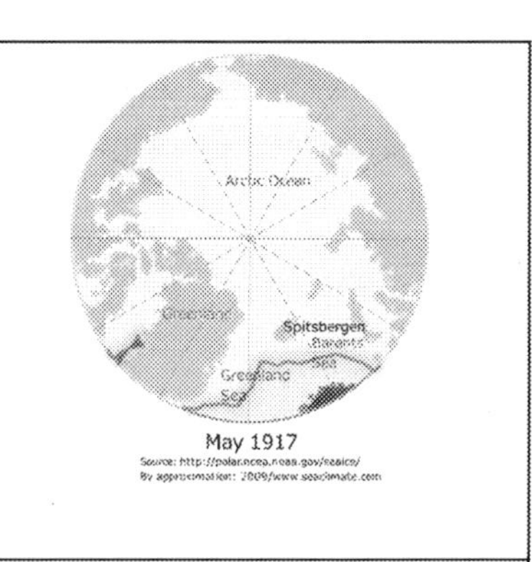

May 1917
Source: http://polar.ncep.noaa.gov/seaice/
By approximation: 2009/www.seaclimate.com

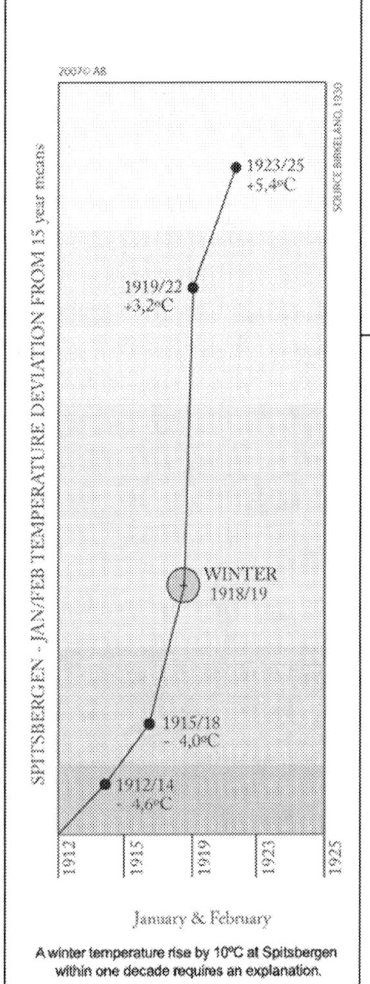

A winter temperature rise by 10°C at Spitsbergen within one decade requires an explanation.

The year with the most southern edge of sea ice was in 1917

with no ice free sea of the Spitsbergen's coast.

The only time during the entire 20th Century.

Arctic Temperature Differences during Winters 1921-1930 minus those for 1911/20

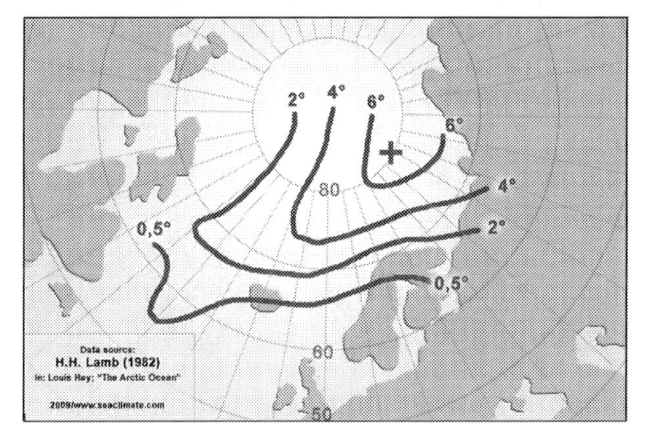

C. What to make out of the big rise?

On one hand there is an extraordinary and very sudden rise in air temperatures at Spitsbergen in the late 1910s, and the more time passes the less is the scientific community showing interest to scrutinize this event thoroughly and to search for clues. That is too little and insufficient. It appears that the generations of the scientist in the 1930s were more willing to take a comprehensive approach in Arctic climatic matters, and more willing making progress in this respect, as demonstrated by a sentence made many decades ago: "It has for some years surprised me that although the motions of the oceans of air and water have much in common and depend on the same principles, students in meteorology do not as a matter of course acquaint themselves with the fundamental facts of oceanography" (Carruthers, 1941).

The book "Oceanography for Meteorologists" published in 1942 (Sverdrup, 1942) [27]already stressed:

> *"It might appear, therefore, as if the oceanic circulation and the distribution of temperature and salinity in the oceans are caused by the atmospheric processes, but such a conclusion would be erroneous, because the energy that maintains the atmospheric circulation is to a great extent supplied by the ocean.*

The following discussion will pay attention to the advice. Having established that the temperature showed a rocket rise in winter 1918/19 and remained significantly high over two decades until 1940, this solid fact needs now to be analysed in its wider context and how it could be generated and sustained.

[27] See also: Special Page "2005, Polyakov", at Chapter 4

1938

Overland, J.E.(2008) & M. Wang, S. Salo;
„The recent Arctic warm period";
Tellus (2008), 60A, p. 589–597

Abstract (Extract): A meridional pattern was also seen in the late 1930s with anomalous winter (DJFM) SAT, at Spitsbergen, of greater than +4°C. Both periods suggest natural atmospheric advective contributions to the hot spots with regional loss of sea ice. Recent warm SAT anomalies in autumn are consistent with climate model projections in response to summer reductions in sea ice extent. The recent dramatic loss of Arctic sea ice appears to be due to a combination of a global warming signal and fortuitous phasing of intrinsic climate patterns.

3. Northern Hemisphere climate patterns (Extract): The only major departure in the 20th century was during the 1930s when SAT observations at Spitsbergen had an extended interval with winter (DJFM) anomalies above +4°C relative to a 1912–2002 baseline (Fig. 7a). Maximum temperatures were toward the end of the decade with composite SLP (sea level pressure) anomalies for winter 1937–1939 showing strong meridional flow towards Svalbard.

(NOTE: Fig.7a indicates a temperature difference between 1917 and 1920 of 11°C. The mean level remained zero until 1925, turned to about –3° until 1930, to continue on a ca. mean level of +2° until WWII.)

6. Conclusions (Extract): The SAT and SLP patterns in the central Arctic at the beginning of the 21st century (2000–2007) were unique compared with most of the 20th century and are labeled the Arctic warm period.

The winter/spring SLP anomalies for 2000–2007 often have a pressure dipole/meridional geostrophic wind pattern with some resemblance, but different orientation, to the pattern in the 1930s, when the AO (Arctic Oscillation) and PNA* (Pacific North American-like) were also small.

Question:

Why is no attention paid to the fact that the warming occurred more than 10 years earlier than the researcher mention, namely "during the 1930s" as they did also in previous work. See: Overland, J.E. (2005) and Muyin Wang; 'The third Arctic climate pattern: 1930s and early 2000s', when saying: "The period from 1928–1935 also had a dipole structure in SLP, which contributed to the interdecadal arctic-wide warm temperature anomalies in the first half of the 20th century."

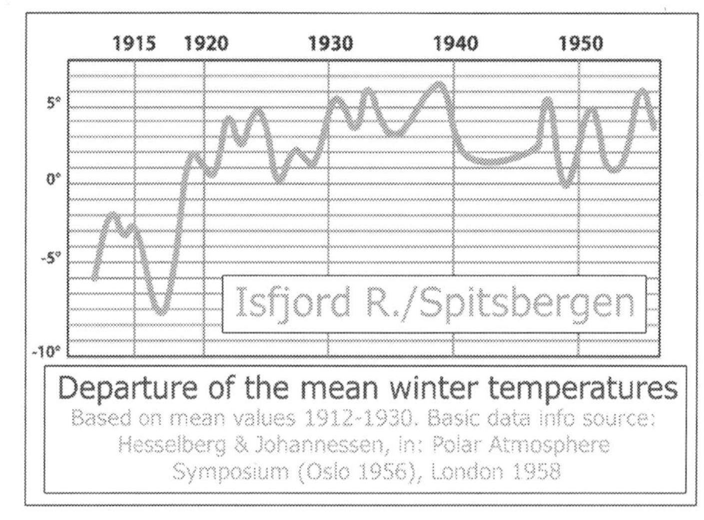

Departure of the mean winter temperatures
Based on mean values 1912-1930. Basic data info source: Hesselberg & Johannessen, in: Polar Atmosphere Symposium (Oslo 1956), London 1958

Chapter 4
Regional impact of the Spitsbergen warming

A. Introduction to When & Where

Being impressed by the big warming at Spitsbergen is one thing, another one, to make sense of it by putting the sudden temperature increase in winter 1918/19 in a perspective. How was the event recorded in other locations of the wider region? At what time could a commencement of the warming be observed, was it before Spitsbergen or after Spitsbergen; was it more pronounced or less pronounced, and how long did it last, two decades unit 1940 as in Spitsbergen or ended the warming up trend earlier? In so far this chapter can be called a fact finding mission. It aims at preparing the basis required to discuss matters of correlation and causation. After all, the heat that increased the temperatures in the Arctic region must have been generated either locally or somewhere else. If generated outside the Polar Circle, it should be possible to identify the generating source and the path of transport either through the atmosphere or the ocean system.

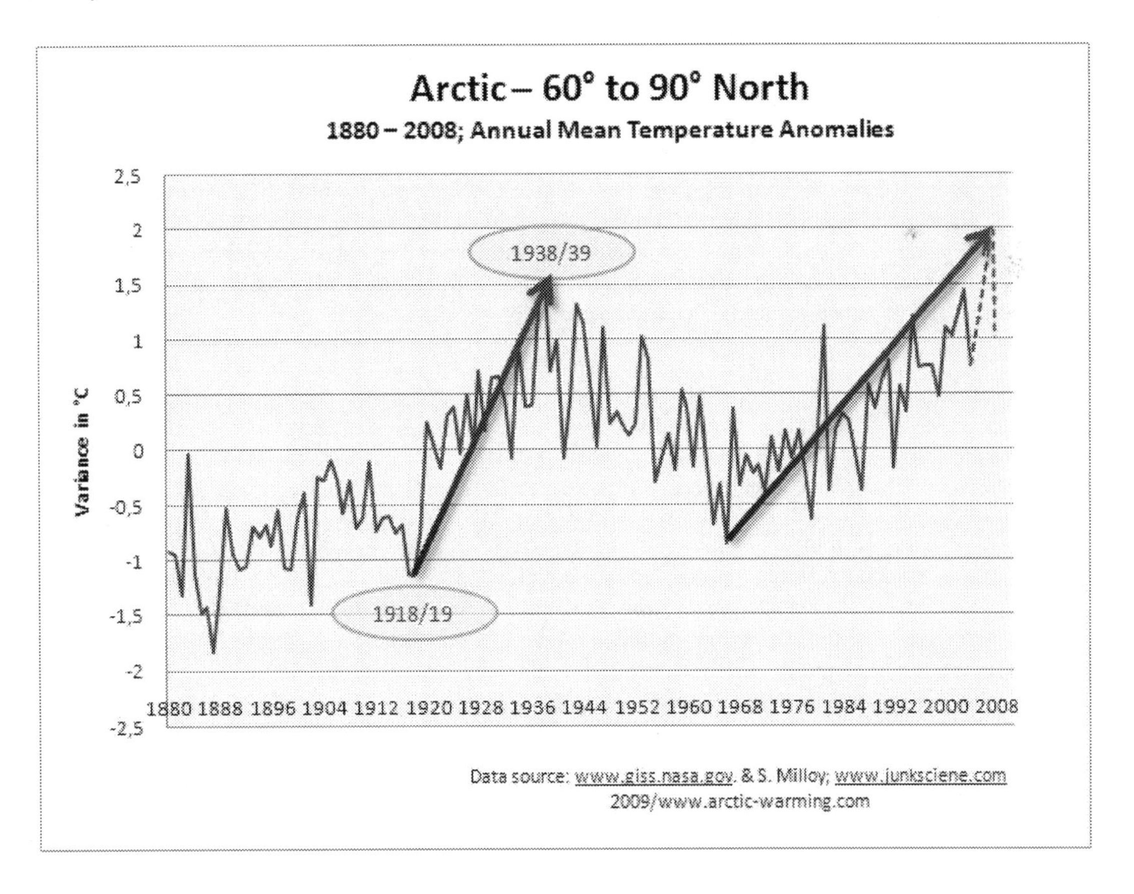

4. Spitsbergen temperature rocketing

When discussing "the nature of polar forcing" (Zakharov, 1997), the author stressed that the feature of climate change spatial structure in the 20th Century is most well pronounced in the northernmost part of the Atlantic and in the adjacent Arctic regions, which supports the idea that obviously, the atmosphere thermal regime is in some way connected with the ocean. To support this idea, as we will do in the next chapter, a broader picture on facts and information portraying the Spitsbergen event is required.

B. The regional feature with four sea water bodies

The availability of large water-bodies so high in the Northern Hemisphere are the ultimate "blueprint" for the Northern Hemisphere climate, which needs to be recognised in the first place. However such plain statement cannot explain the "climate revolution" in the region in the early 20th Century. Calling this initial event "Arctic Warming" might even be misleading, if the warming commenced primarily over the northern part of the North Atlantic, and even more precisely, in the Spitsbergen region. On the other hand all ocean space and seas within the Arctic Circle are relevant for the arctic weather and arctic climate, although on a very different scale due to size, depths, sea ice, freshwater, sun-less periods, and sea currents, to mention only few decisive factors. In order to detect the time and source of the sudden start of the Arctic Warming one must have a look at each of the specific sea areas in question. This needs to be kept in mind that the "explosion" of air temperatures at the Spitsbergen archipelago at 80° North during winter 1918/19, must have been generated by "something".

Although no option, including an internal atmospheric process, should be excluded, the most likely source is the sea area between the direction 135° (SE) and 270° (West) of Spitsbergen, which is usually sea ice-free throughout the year and belongs to the Barents Sea, the Norwegian Sea and the Greenland Sea. The source of the warming was presumably due to either internal processes within the water bodies, or influenced by more warm water coming from the Atlantic Gulf current, or both. The latter came with the Norwegian Current and West Spitsbergen Current, formed by water flowing from the Gulf Current after it had passed the Iceland - Faroe – Scotland line, enhanced by North Sea water, and continental run off rain and melt water. However, once the warm Atlantic water has passed the north of the Scottish Hebrides and Faroe, and travelling northwards, things tend to complicate.

a) Arctic Ocean

Due to the size of the Arctic Ocean, which is much larger than the Northern North Atlantic[28], and due to its considerable depths of more than 3000 metres, one would assume that its immediate impact on the weather should succeed by far the influence of the Northern North Atlantic. That is not the case. At least during the winter periods in the early 20th Century the Arctic Ocean was permanently covered by sea ice, which diminishes any heat release from the sea in the atmosphere to a small factor, thus making this huge water body "continental", characterized by cold and stable weather conditions, and clear skies. However the sea ice is not very thick, and not always unbroken, so that an interchange is not completely excluded. According to Aagaard (Aagaard, 1982) it seems possible that two-third of the oceanic flux should be accomplished by the Atlantic water of the West Spitsbergen Current.

The feature of the Arctic Ocean is very complex and only few aspects can be mentioned. Above the very cold and saline bottom water is a layer of "Atlantic Water", which occupies the depth range between ca.

[28] Here meant as area between: Greenland, Iceland, Scotland, Norway, and Spitsbergen.

4. Spitsbergen temperature rocketing

Annual Air Pressure deviation 1921-1930

H +2mb

H +0,5mb

H +0,5mb

ca. -1,4 to -2,0 mb

L

Data source: R. Scherhag (1936) by approximation 2009/www.seaclimate.com

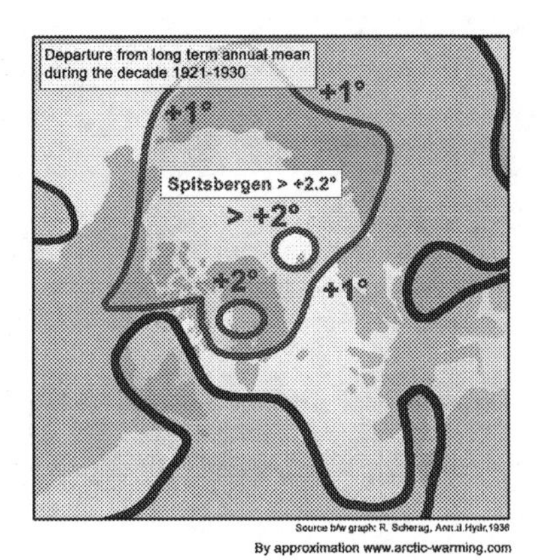

Source b/w graph: R. Scherag, Ann.d.Hydr,1936

By approximation www.arctic-warming.com

Deviation from mean 1921-1930, air pressure (left), air temperature (right)

Basins and Shelfs in the Arctic Region

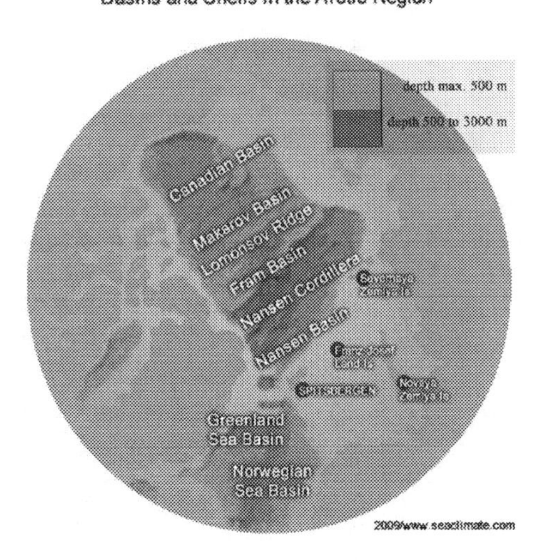

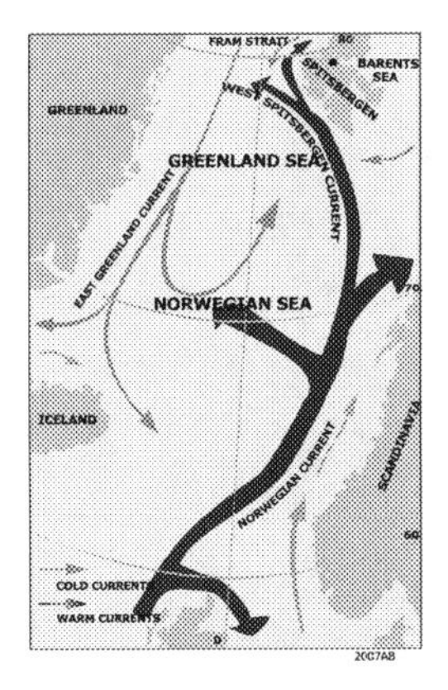

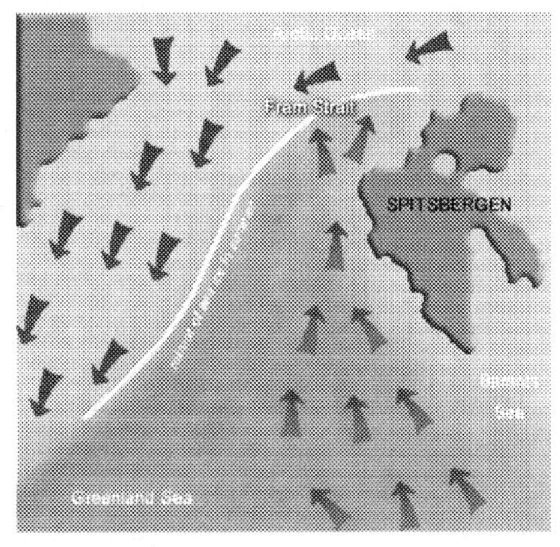

1936

Schokalsky, J. (1936)[1] ;

"Recent Russian researches in the Arctic Sea and the in mountains of Central Asia",

in: The Scottish Geographical Magazine, Vol. 52, No.2, March 1936, p. 73-84.

Extract from the paper:

____This work (of Russian scientists and institutions) was necessary in order to ascertain the temperature of the Atlantic branch of the Gulf Stream west of Spitsbergen, and to know what temperatures conditions may be expected along the Eurasian continental slope of the North Polar Ocean.

____During the memorable voyage of the FRAM (1893-1896), Nansen discovered that the upper layer of the Arctic Ocean from 200 to 500 metres in thickness was less saline than the deeper water, and that it had a temperature of –1,0° to –1,9° C, while the deeper layer, from 600 to 700 metres tick, was of oceanic salinity (over 35 per cent.), and had a temperature of +1,2°C.

---------- Five years later, S.O. Makazov, on the ice-breaker ERMAK, between Franz Josef Land and Novaya Zemlya, found that the zero temperature occurred at a depth of about 200 metres, and that below this the temperatures rose to +1,1°C. This confirmed Nansen's observations.

----------- In 1927 the ship ELDING, between Franz Josef Land and Novaya Zemlya, recorded temperatures of +0,6°C at 100 metres depth. Then, in 1928, the KRASSIN'S observation north of Spitsbergen in lat. 81°47'N revealed a new fact. At 70 metres the water was found to have oceanic salinity of over 35 per cent and a temperature of –4,6°C.

----------- Sverdrup's observations in the NAUTILUS in 1928 confirmed these facts for latitude 82°N. In 1929 the SEDOV and the PERSÉE, at almost the same place Makazov chose for his observation in 1901, found the zero isotherm to occur at the depth of 125 metres instead of 200 metres.
----------- Again, in 1931, the PERSÉE, in the same vicinity, found this thermobath at 75 metres, and the LOMOHOSOV, a little farther eat, found it from 25 to 40 metres, below which depth the temperature increased to +1,6°C. (cont.)

----------- And finally the PERSÉE in 1934 reaching 81°17'N, north of Spitsbergen, early in September, recorded an air temperature of 12°C and a sea temperature of +5,5°C down to 10 metres.

____The branch of the North Atlantic Current which enters it by way of the edge of the continental shelf around Spitsbergen has evidently been increased in volume, and has introduced a body of warm water so great, that the surface layer of cold water which was 200 metres tick in Nansen's time, has now been reduced to less than 100 metres in thickness.

____These records, and others not cited here, together provide incontestable evidence of a progressive warming of the Arctic Ocean.

____For this purpose, it is necessary to know more about the thermal conditions of the branch of the Atlantic Current which passes round Spitsbergen.

NOTE: In 1953 H.W. Ahlmann provided the information: The thickness of the ice forming annually in the North Polar Sea has diminished from an average of 365 cm at the time of Nansen's Fram expedition of 1893- 96 to 218 cm during the drift of the Russian icebreaker Sedov in 1937-40. (Ahlmann, 1953).

---

[1] Professor Jules Schokalsky , President of the Geographical Society of the Soviet Union. A paper read before the Royal Scottish Geographical Society in Edinburgh on the 30th January 1935, on the occasion of the presentation to Prof. Schokalsky of the Society's Research Medal.

150m and 900m that is superseded by the "Arctic Surface Water" in the range from the sea surface to a depth of 150/200m and temperatures close to the freezing point (between –1.5° and –1.9°C). In this layer the salinity depends strongly on the degree of the sea ice processing, freezing or melting, ranging from about 28 to 33.5°/o. These conditions represent in the Arctic a relatively warm ocean surface water layer which has an influence on the regional air temperatures.

Despite the fact that the water body of Arctic Oceans is based on many complex features, and highly influenced by internal transformation, e.g. sea ice freezing and melting, and external influences, e.g. fresh water from rivers, there is not one aspect from which a sudden warming could have been generated. None of the seas in question can be excluded with such unequivocal certainty from the list of potential contributor for the early Arctic Warming as the Arctic Ocean.

This assertion is addressing only the sudden warming in the late 1910s. The huge water body in an extreme harsh environment will transfer any internal changes ultimately to the atmosphere, presumably with considerable time lags. One needs only to consider the water renewal time for the Arctic Bottom water, which is a couple of dozen years in the Amundsen- and Nansen-Basin but more than 500 years in the Canada- and Makarov-Basin. Due to the long internal processing most recent dramatic arctic warming and sea ice melting may have been caused by warm Atlantic water that arrived many decades ago. In order to find out, it would primarily require establishing what and why the Spitsbergen warming occurred in the first place, as it constitutes the early Arctic Warming from 1918 to 1940.

b) Greenland Sea

A significant climatic impact is due to the southward flowing East Greenland Current (EGC) which constitutes the major outflow route of Arctic water into the Atlantic. The pathway to the south of Greenland is due to the season covered with sea ice. The water temperature is below –1°C, and due to ice melting of low salinity of about 30-33 °/o. Some of it is diverted just north of Denmark Strait and northeast of Iceland into the East Iceland Current, which before reaching the latitude if Iceland towards the Norwegian Sea as part of the formation process of Arctic Bottom Water. The remainder pass the Denmark Strait, gets partly mixed[29] with the Irminger Current and in this combination flow around the southern tip of Greenland, Kapp Farvel, into the West Greenland Current[30]. Cold sea surface temperature of the EGC is associated with negative anomalies of surface air temperatures with an amplitude of 2° near Greenland declining to several tenth of a degree over north-western Europe (Delworth, 1997). In the west of Spitsbergen, the seawater has a temperature of 5°C and a salinity of 34.90 –35.00 mg.

A significant part of the warm Atlantic Gulf water that has reached Spitsbergen "turns left" in the south-western direction, at the position of 75-77° North, and flows either as Greenland current down to Newfoundland and back in the Atlantic, or goes down into the huge Greenland Sea Basin with depths of 2,000 metres (max. ca. 3,500 m), or circles for some time at the sea surface water layer or in sub surface layers. At the surface the water forms a layer of 100 –200m, which reflects the depth of summer water

[29] Transport estimates are 5 Sv (5 × 106 m3 s−1) for the East Greenland Current and 8–11 Sv (8–11 × 106 m3 s−1) for the Irminger Current. The combined flow continues around the southern tip of Greenland into the West Greenland Current.
[30] The Labrador Current originates from East Greenland Current, continues as West Greenland Current (NW), than as Baffin Island Current (SW), and subsequently the Labrador Current (SW), which transports cold Arctic and sub-polar water south along the Atlantic Canadian coast to the Grand Banks (Newfoundland) where it divides, the East branch joining the North Atlantic Current and the West branch flowing into the Gulf of St. Lawrence. The International Ice Patrol had first quantitatively studied the Labrador Current in 1937.

heating. Within the annual cycle, these water body experience considerable deviation of temperature (4°C) and salinity (1.2 °/o).

Although the Greenland Sea represents a huge water body, the option for being a serious contributor to the extreme warming event in winter 1918/19 is remote. The West Greenland Current is colder and less salty than the Atlantic water of the Spitsbergen Current, and the current is to a considerable extent covered with sea ice during the winter season, but only partly in summer. Concerning this investigation the Greenland Sea matter gets "hot" if one looks at the EGC and whether the current had been getting more than usual portion of warm Atlantic water that made the temperatures exploding at Spitsbergen around winter 1918/19. Did the thus warmed up EGC current brought a warming to Greenland"s coast in the East and the West of the island? We will discuss this aspect later, while mentioning already now, that Greenland had experienced warmer sea water and warmer air during the time period in question, but later and only temporarily (Bjerkness, 1959).

c) Norwegian Sea

The Norwegian Sea Basin is up to ca. 3,000 metres deep, with a mean depth of 1740 m. The sea is separated from the Arctic Ocean by a ridge of about 600 metres under the sea surface. This allows huge water exchange from north to south and vice versa. A key role for the moderate temperature conditions in the region derives from huge warm water supply by a branch of the Atlantic Gulf currents which is "pushed down" to lower depths after passing by the Shetland Islands, Faroe Island and Iceland ridge, travelling North about 500 m below the sea surface. On the first stretch parallel the coast line of Norway, the warm Atlantic water is called Norwegian Current, which is to be kept separate from the Norwegian Coastal Current, but once it has reached the latitude of the North Cap the major part continues as West Spitsbergen Current (WSC) towards the Fram Strait in the West of Spitsbergen, and a small part goes eastwards in the Barents Sea. Due to its high northern location, size and volume, the Norwegian Sea has a lot of facets that are too complex for being presented in few words.

If for example the wind would not mix a low saline sea surface, due to rain fall, with saltier water from lower sea levels, the sea ice would presumably reach Iceland, Scotland and the North Sea at least every winter season. That would even happen when the most unique feature, the West Spitsbergen Current, would pass the Norwegian Sea as usually, but latest demonstrate its impact on the Arctic climatology as soon as the current has arrived in the West of Spitsbergen. This is of foremost interest, and any other details concerning the Norwegian Sea will be mentioned if deemed helpful.

As already mentioned, the WSC is the major source of heat and salt for the Arctic Ocean. The Current cools dramatically downstream, although the warmest water can be frequently observed 100-200m below the surface moves via the currents towards the basin of the Arctic Ocean. For the last distance over ca. 1000 km, Lofoten to Spitsbergen the travel time is about 5 weeks, which means the average speed is 0.30 to 0.35ms^{-1} Actually, due to the high salinity of the warm Atlantic water and the cooling process, the water becomes very dense and "falls" over a ridge (with a depth of 600 m below sea level) in the Arctic Basin. Before the Spitsbergen current reaches the ridge, at about 80° North, the water, has a depth of 20 metres, a salinity of about 35 per mile, and a temperature > 3° up to 7°C.

But also the basin of the Norwegian Sea is a reservoir for warm Gulf water, reaching depths of 800 metres. Status and dynamics of the Norwegian Sea is also strongly influenced by other factors, particularly wind, rain, melt water, and the low saline water form the North Sea. Any increase in temperature, or enlargement of the "warm water part", or "change of dynamics", would quickly be reflected in temperatures in Europe or elsewhere in the Northern Hemisphere.

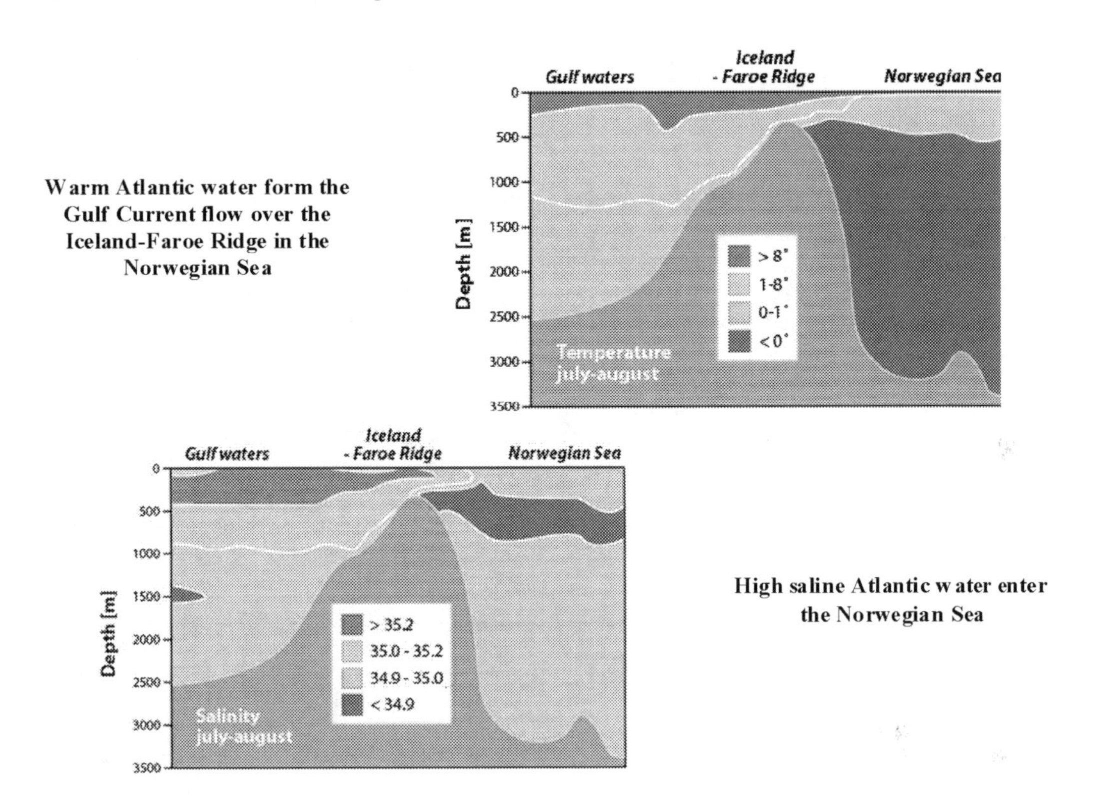

Warm Atlantic water form the Gulf Current flow over the Iceland-Faroe Ridge in the Norwegian Sea

High saline Atlantic water enter the Norwegian Sea

d) Barents Sea

The Barents Sea is a shallow shelf sea with an average depth of 230m, and a maximum depth of below 500m. There are three main types of water mass; warm, salty Atlantic water (temperature >3°C), cold Arctic water, less salty (< 0°C), and warm coastal water, less salty than the Arctic water (>3°C).

Generally speaking, the warm Atlantic water "disappears" in the East of the North Cape. In the northern part polar water flows from Northeast to Southwest and partly joins the Spitsbergen Current in the south of Spitsbergen and Bear Island. The North Cape Current, and its subsequent currents, which supply the eastern part of the Barents Sea with Atlantic water, may have contributed to the warming in the long run, under the condition that a permanent inflow of warm Atlantic water is guaranteed. Actually the whole water body of the Barents Sea is replaced within a 3 – 4 year period. The Barents Sea itself is not able to sustain a longer warming period due to limited heat storage.

With regard to the Spitsbergen, it has already be mentioned that the air temperatures in North Norway have increased only modestly since 1920 (Manley; 1944). Neither Vardö (close to the North Cape), nor any Russian Station reported any exceptional winter temperature rise. On a 10-year mean basis (1911/20 and 1921/30), a significant increase of 6°C was observed at Franz-Joseph Land (Kirch, 1966), whereby the highest water temperatures in the top 200 m of the Barents Sea at 70-72°N; 33°E, north of the Kola Peninsula, shall have been reached during the period 1935-39 (Lamb,

2005	Igor V. Polyakov et al. (2005); **"One more step toward a warmer Arctic"** Geophysical Research Letters, Vol. 32, L17605; Co-authors see: Reference list

The ABSTRACT notes that the study was motivated by a strong warming signal in2004 in the Eurasian Basin of the Arctic Ocean, claiming, that the source of this and earlier Arctic Ocean changes lies in interactions between polar and sub-polar basins. Evidence suggest such changes are abrupt, or pulse-like, taking the form of propagating anomalies that can be traced to higher-latitudes. For example, an anomaly found in 2004 in the eastern Eurasian Basin took _1.5 years to propagate from the Norwegian Sea to the Fram Strait region, and additional _4.5–5 years to reach the Laptev Sea slope.

The paper furthermore states:

- This evolution of water temperature is related to the atmospheric processes: (p: 2)
- The abrupt nature of these warming events is striking. The first temperature increase at the EEB slope of about 0.4_C in February 2004 happened in a single day, after which the AW layer equilibrated at a new warmer state for almost seven months, when another abrupt warming occurred. (p: 3)
- Through analysis of a vast collection of observational data it was shown that over the 20th century multi-decadal AW fluctuations are a dominant mode of variability (Figure 3). Associated with this variability, the AW temperature record shows two warmer periods in the 1930–40s and in recent decades, and two colder periods early in the 20th century and in the 1960–70s. (Concluding Remark)

Subsequently Igor A. Dmitrenko, Igor V. Polyakov , et al (2008) acknowledged: "recent Atlantic Water (AW) warming along the Siberian continental margin due to several AW (Atlantic Water) warm impulses that penetrated into the Arctic Ocean through Fram Strait in 1999–2000". One year earlier The New York Times (2nd October 2007) – by Andrew C. Revkin - referred to Igor V. Polyakov, saying that he see a role in rising flows of warm water entering the Arctic Ocean through the Bering Strait between Alaska and Russia, and in deep currents running north from the Atlantic Ocean near Scandinavia.

Comment:

The stunning point is the statement that the "evolution of water temperature is related to the atmospheric process", despite the various other mentioned aspects whereby the sea water records indicate to propagate anomalies. Do I.V. Polyakov and colleagues really think that the branch of the Gulf Current, the warm Atlantic Water on its way to Spitsbergen, is depended on atmospheric processes? Have they ever evaluated their conclusion in conjunction with the winter sea ice? At least concerning the early Arctic warming? What else than the warm Norwegian- and Spitsbergen Current could suddenly have initiated and sustained the sudden warming of the Arctic winter seasons. Climate is the continuation of the oceans by other means, namely heat & vapor. One of the best scientist in this field some time ago would presumably have advised to read the following from his book: "Oceanography for Meteorologists" published in 1942:

> *It might appear, therefore, as if the oceanic circulation and the distribution of temperature and salinity in the oceans are caused by the atmospheric processes, but such a conclusion would be erroneous, because the energy that maintains the atmospheric circulation is to a great extent supplied by the ocean. It will be shown that this energy supply is very localized, owing to the character of the ocean currents, and that therefore the circulation of the atmosphere, which depends upon where energy is supplied, must be influenced by the oceanic circulation. The reasoning leads to the conclusion that one cannot deal independently with the atmosphere or the oceans, but must deal with the complete system, atmosphere-oceans. This fact has been recognized in oceanography, where one gets nowhere by neglecting the relation to the atmosphere, but in meteorology it has not yet received sufficient attention. (p. 223) It is reasoned that the heat content of the ocean water is very great compared to that of the atmosphere, and that therefore any change in ocean current will for a long time influence the air temperature and the circulation. (p. 234)*
>
> H.U. Sverdrup, 1942, "Oceanography for Meteorologists", New York, 1942, Chapter X, p. 223

1980). That all confirms a slow process but hardly any facts that the Barents Sea had significantly generated a temperature jump in its southern part (North Cape), or in the eastern part (Kola Peninsula).

The observed warming at Franz-Joseph-Land can also be hardly connected to the sea, which is partly supplied via the North Cape current and subsequent currents along the Kola Peninsula. In addition the Barents Sea between Spitsbergen (East) and Franz-Joseph-Land (West) is not very deep and is governed by very cold currents flowing southwest towards South Spitsbergen and the Bear Island. (See also next section).

According to Wagner (Wagner, 1940), the mean surface-water temperatures in the Barents Sea, the deviations from the means during the months July to September, was –0.7°C during 1912-1918, against + 1.1°C during the following years 1919-1928, as indicated in the table:

1914 = -0.3°C	1915 = + 0.7°C	1916 = -1.1°C	1917 = -1.5°C	1918 = -1.6°C
1919 = + 0.6°C	1920 = +2.2°C	1921 = +1.0°C	1922 = +1.9°C	1923= +1.0°C

Wagner's additional observations ascertain a "rise" of 2-3°C, at water depths of 100 and 200 m, during the last 30 years (1895 and 1927). However, a general observation of "over 30 years" is of little help in this case, presumably also his general assessment that the Barents Sea ice border retreated significantly since 1919, even though it is undisputed that the retreat occurred gradually over a two-decade period (Kirch, 1966).

There are no significant indications that the Barents Sea contributed significantly in the early stage of the Arctic warming. The presumably most relevant aspect is that no particular strong winter warming had been observed at the southern parts of the Barents Sea, albeit a trend change was observed since 1919. The question would remain, whether the trend change had been caused by higher temperatures in the North, or alone by the Barents Sea, or was due to changes over the European Continent, or had been a combination of all three factors. But this does not need to be answered here.

C. Summary

The brief overview on the possible potential of the most relevant sea areas for the early Arctic Warming could show that two, out of the discussed four areas, can be excluded with high certainty as serious contributor, namely the Arctic Ocean and the Greenland Sea, and with less certainty the Barents Sea. The highest potential by far has the northern part of the Norwegian Sea and the Spitsbergen Current. This preliminarily assessment shall be verified with further information in the next section.

4. Spitsbergen temperature rocketing

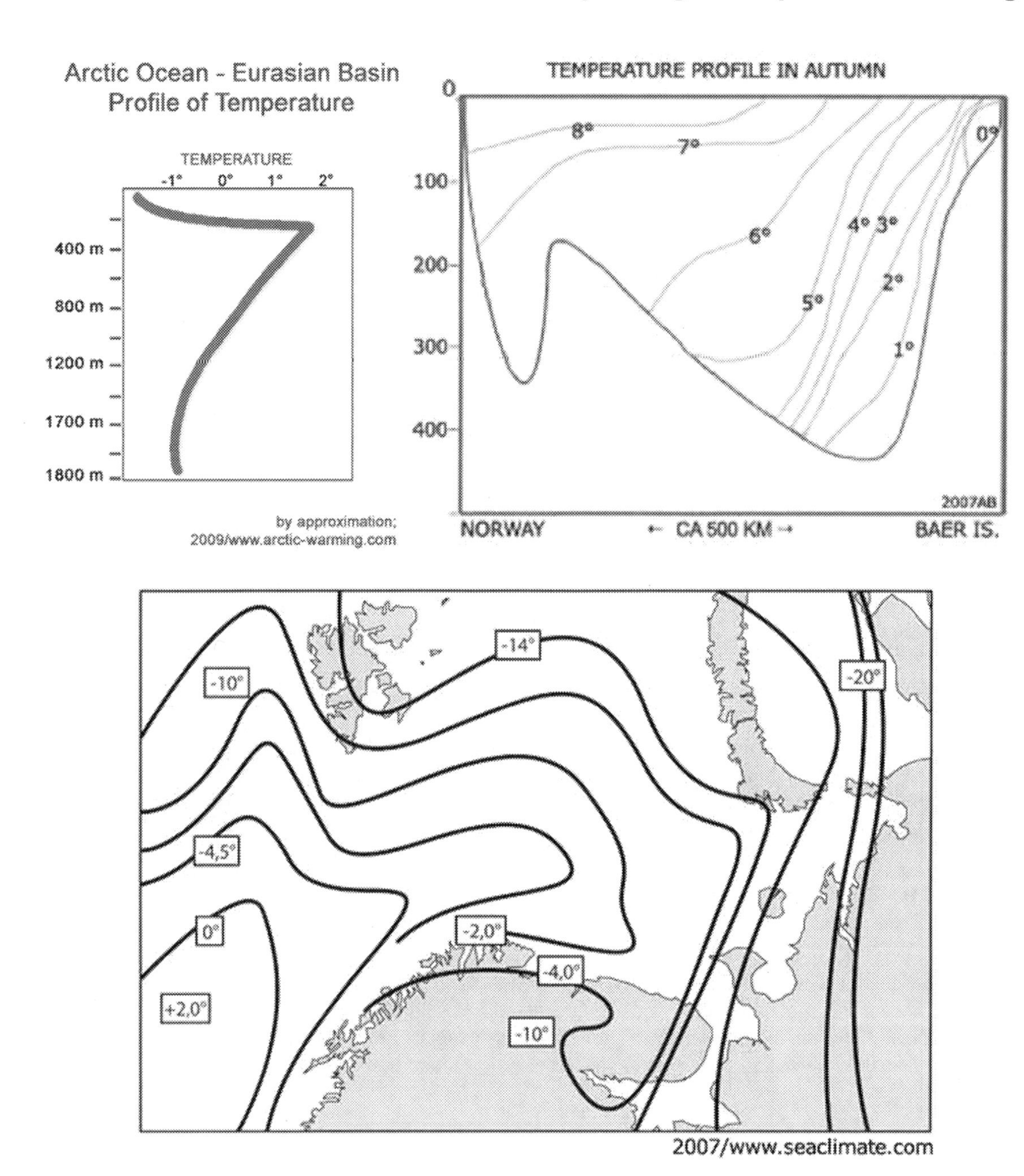

Barents Sea - Ca. Mean January Temperature in °C

Chapter 5
The warming event in details

A. Exceptional temperatures rise

It is said that the Arctic temperatures in the 20th century have been the highest over the past 400 years (Overpeck, 1997). The fact is that until now the global surface air temperature (SAT) have increased by 0.76ºC since 1861[31]. This was not a homogenous rise. The air temperatures through the latter part of the 19th century and the early 20th century were relatively cool compared to years since the 1920s, especially after 1925 (Drinkwater, 2006). The increase in the Arctic was two to three times higher then the global mean, particularly during the two decades from 1920 to 1940, whereby the period from 1918 to1922 displayed an exceptionally rapid winter warming in the circum-Arctic region (Polyakov, 2003). This is emphasized by the observation that annual temperatures from 1920–1940 rose even more markedly than during the post 1970s period (Serreze, 2000).

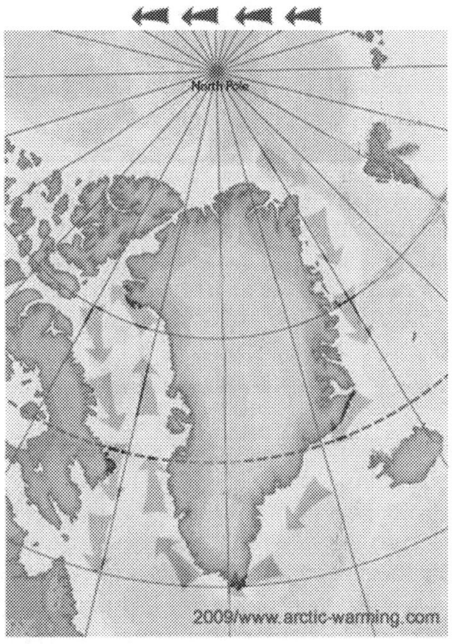

The proportions of the warming are illustrated in almost every Arctic temperature graphic. The moment the event started to happen can be identified very precisely, January 1919. It was the second month after WWI had ended. Suddenly, Spitsbergen a remote archipelago between the North Cape in Norway and the North Pole had corresponding mean temperatures during January as 2000 km further south in Oslo, mere –5°C. That was a climatic "bang". It meant that the temperature differences between the two pre and post WWI January were 16° (sixteen) degrees[32]. Nothing can illustrate the extraordinary jump better than looking at a 10°C difference between the winter temperatures (the mean of D/J/F) seven years before January 1919 (1912-1918), and the seven winters 1919 to 1925. The ability to establish a precise timing is elementary for improving the chance to consider and to reason the circumstances of the event.

Confining the date of the warming-up period to the year 1918 is very precise and important, while describing the warming-up area to "the circum-Arctic region" is rather superficial. On the other hand V.F. Zakharov stated with reference to Russian researcher: "Indicating that while high latitude play a special role in the climate changes over the hemisphere, the Atlantic sector plays a special role in these latitudes

---

[31] IPCC, 2007, The total temperature increase from 1850 – 1899 to 2001 – 2005 is 0.76 [0.57 to 0.95]°C (Summary for Policymaker, p.4)

[32] The January temperature at Spitsbergen : 1917 -20.4; 1918-24.4; (mean: 22.4°C); and 1919 -5.7, 1920 -10.5 (mean: -8.1°C); http://data.giss.nasa.gov/work/gistemp/STATIONS//tmp.634010050010.1.1/station.txt

themselves. This important aspect of spatial structure of polar forcing should not, of course, be omitted when explaining the causes of the present day climate changes" (Zakharov, 1997). This process became pronounced affecting the ecosystem with a general northward movement of fish, due to a dramatic warming of the northern North Atlantic Ocean during the 1920s and 1930s (Drinkwater, 2006). Quite a number of research papers on fishing have mentioned this aspect since the 1930s (e.g. Carruthers, 1941) but remain unspecific concerning details. This is of not too much help if one wants an explanation: what made the northern North Atlantic warming? Meanwhile we have established that Spitsbergen was a hot-spot in this scenario. What we not know yet: Had there been several hot-spots? Became they effective simultaneously or one after the other? Do distant locations and periods of warming or cooling allow conclusions? The starting point is that the changes started in the late 1910s and evidently extreme pronounced at Spitsbergen, and for this reason the northern North Atlantic will be the focal point in our further investigation.

B. Distant warming

Overall, the Earth gains heat at low latitudes by the sun and it loses heat at high latitudes. To balance these gains and losses the heat is transported poleward by the atmosphere and the ocean in comparable proportions. In winter 1918/19 and from thereon a lot of surplus of heat must have been coming from distant places, so that distant and medium of transport matters. This shall be indicated by the term 'distant warming', whereby a region or location which is separated form the warming hot-spot Spitsbergen, is within the "reach" of Spitsbergen either by atmospheric air masses, or particularly by the ocean current system. This means practically that all seas and their coastal areas north or in the range of the Arctic Circle could have contributed by 'distant warming'.

By the 'core Arctic' region shall here be understood the Arctic Ocean, respectively the sea space north of 80° North and north of the all continents[33]. In addition to the exclusion of the landmasses, it seems that also the Northern Pacific Ocean region can be excluded. Neither has any research in the 1930s, or more recently asserted that the early Arctic warming could have been generated in the Alaska, Siberia and North Pacific sector of the Arctic Ocean. As the Aleutian in the North Pacific are at the latitude of 51° north, that is on the same latitude as London, Kiev, Astana, and Vancouver, the Bering Strait region is not comparable with the situation in the Northern North Atlantic.

The very different oceanic conditions between the two ocean entrances to the Polar Basin are even more significant. The fairly cold and low saline water inflow from North Pacific to the Arctic Ocean is extremely less influential than the warm water masses, which arrive via the Norwegian and Spitsbergen Current at Spitsbergen and flow into the Arctic Basin. The volume of Atlantic water is about eight times higher then the mass transport via the Bering Strait.

C. Is ocean warming ascertainable?

It would be so helpful if the "most significant regime shift experienced in the North Atlantic in the 20th century"(Drinkwater, 2006) could be analysed with data records from all sea levels ranging from the sea

[33] According the International Hydrographic Organization the Arctic Ocean proper extents well over land and other sea areas, see: http://en.wikipedia.org/wiki/Arctic_Ocean#Geography

1995 UNEP/GRIDA: World ocean thermohaline circulation.

The global conveyor belt thermohaline circulation is driven primarily by the formation and sinking of deep water (from around 1500m to the Antarctic bottom water overlying the bottom of the ocean) in the Norwegian Sea. When the strength of the haline forcing increases due to excess precipitation, runoff, or ice melt the conveyor belt will weaken or even shut down. The variability in the strength of the conveyor belt will lead to climate change in Europe and it could also influence in other areas of the global ocean (Source, see below).

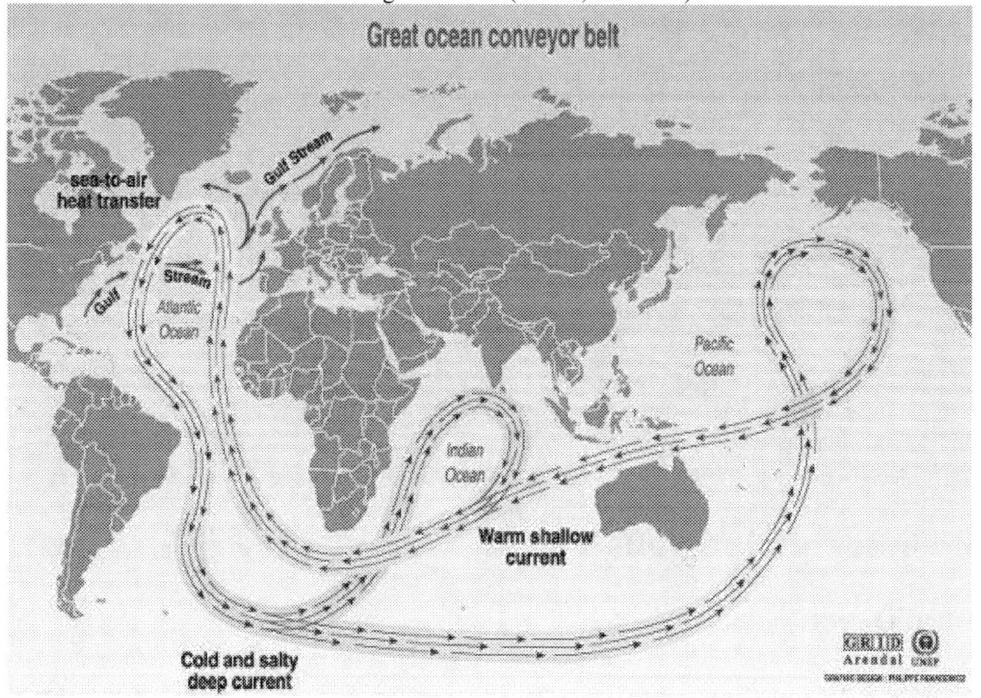

Sources Climate change 1995, Impacts, adaptations and mitigation of climate change: scientific-technical analyses, contribution of working group 2 to the second assessment report of the intergovernmental panel on climate change, UNEP and WMO, Cambridge press university, 1996; http://www.grida.no/publications/vg/climate/page/3085.aspx
Designer Philippe Rekacewicz, UNEP/GRID-Arendal

REMARK

The map identifies only one region as "sea-to-air heat transfer", of which the text actually should be further to the East (close or north of Iceland), and the Gulf Steam symbols should pass Spitsbergen in the West. However, the warm water for the Arctic, from all oceans as far away as the North and South Pacific, is coming with the red line as West Spitsbergen Current to the Arctic gate. As this warm salty water reaches the Fram Strait in the West of Spitsbergen, it is able to release enormous heat into the atmosphere, by cooling down. These cool salty waters are now very dense compared to the surrounding waters, and sink to the bottom of the Arctic Ocean, or Nordic Sea. There is virtually no other place in the world where "heat release" and warming of an cold environment could be more effective. Even a water temperature of only one or two degrees are still able to heat the atmosphere. The end of the squeezing out is only reached when the sea water is freezing. Only this heating-spot seems capable to initiate and sustain an explosive temperature rise, and sustain it over a longer period of time.

surface (SS) to the bottom of the seas. But there had been nothing comparable to air temperature records. Even SST are few and often insufficient, respectively do not exist at all. It is therefore virtually impossible to assess the internal warming processes of the Northern Hemisphere seas from 1918 to 1939.

But the case is not hopeless, as it is possible to draw reasonable conclusions from surface air temperature (SAT) records, which are not overwhelmingly available in number but evidently sufficient to establish a reliable picture of what happened above the ocean surface. This means that it is possible to show a statistical trend of SAT in certain regions of location, but partly – and together with some imagination - what actually happened below the sea surface. In this respect it seems reasonable to listen to an advice given by J. W. Sandström soon after World War II:

> *"The influence of the sea on the weather seems to be of particular interest, seeing that the great specific heat of the sea water influences the atmosphere in a very high degree. The slow variation of the oceanic conditions will probably make it possible to predict the general character of the weather a long time in advance, to foresee the summer in the following winter will be warm or cold, abundant in snow or not, and in the winter if the following summer will be warm and dry or cold and wet." (Sandström, 1947)*

While Sandström wanted to predict weather due to prevailing sea condition, here we are going to analyse the air temperature records in retrospect; which will allow to make conclusions of SST trends. There can be no doubt that any coastal location is considerably influenced by the seasonal and internal conditions of the sea, which is particularly influential at a location as Spitsbergen far away from continental reach.

D. The influential regions

a) Arctic Ocean

During the early years of the last century the Arctic Ocean was almost completely out of reach for the collection of meteorological data series. Neither SST nor SAT records exist. The situation is furthermore complicated by the extensive availability of sea ice, and related to seasonal changes. The maritime influence grossly diminishes, the higher the Arctic Ocean sea ice cover is. Any sea ice free ocean space during the summer season will substantially increase the interchange with the atmosphere (e.g. with regard to its air temperatures, the humidity, and the cyclonic and anti cyclonic conditions). Back in the early part of the last century the meteorology was already well versed with preparing weather charts and drawing climatic relevant conclusions from the material.

There is, on one hand, clear indication that the whole Polar region warmed up over the two decades, 1920s and 1930s , more than any other region anywhere in the world. On the other hand, the information available provides little hindsight on the hypothesis questioning whether the observed Arctic warming is the result of "summarized" temperature statistics on a decadal basis, or whether certain particular warming spots could be identified.

Today it is an acknowledged fact that only the part of the Arctic closely connected to the Northern North Atlantic experienced an exceptional temperature rise (Johannessen, 2004). The most pronounced warming area from 1920-1939 covered a region from the East coast of North Greenland (60° West) to

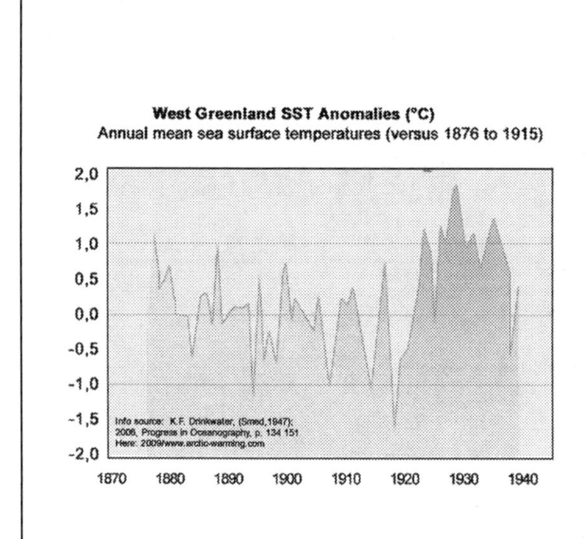

Upernavik - West Greenland

Annual mean temperatures °C

Source: NASA-GISTEMP - http://data.giss.nasa.gov

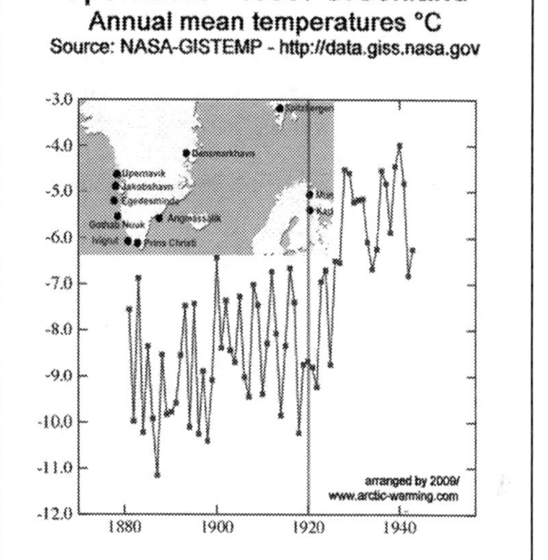

West Greenland – Lowest 1918

Annual mean sea water (SST) anomalies

SW Iceland – Lowest 1921

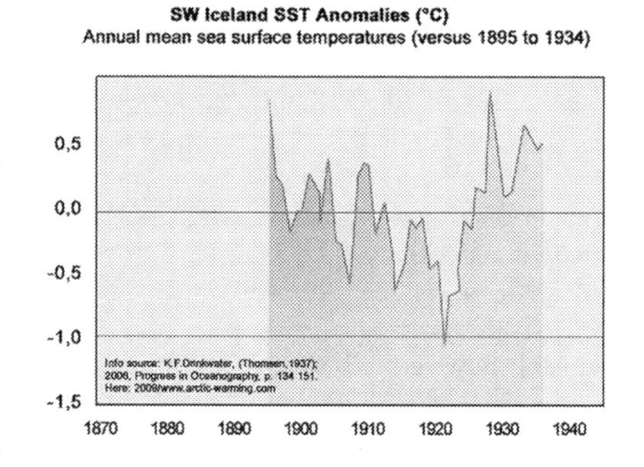

Trend Change after 1920

Annual mean air temperatures

Trend change after 1920

Reykjavik-Iceland

Annual mean temperatures °C

Source: NASA-GISTEMP - http://data.giss.nasa.gov

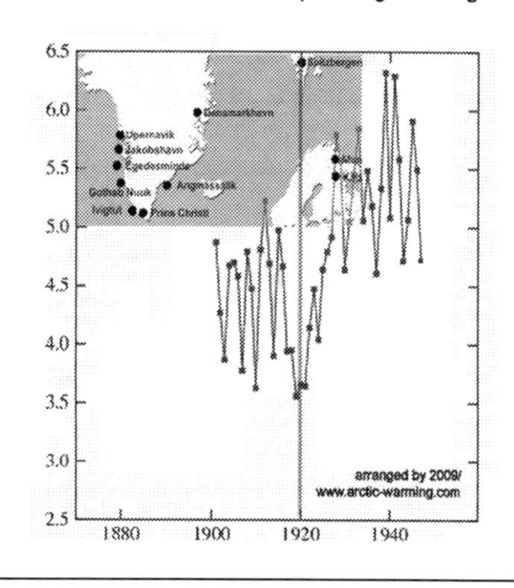

Severnaya Zemlya Island (100° East)[34]. This was already observed since the 1930s by a number of researchers (e.g.Scherhag, 1939), and again confirmed by H.H. Lamp who noted that "the change of prevailing temperatures seems to be the greatest in the regions affected by changes in the balance between the warm northbound Atlantic water and the cold polar current at the ocean surface in the Norwegian-Barents Sea-East Greenland region" (Lamp, 1982). The fact whether Lamb's assertion makes sense and would permit the 'explosion' of temperatures at Spitsbergen will be discussed later. Comparing the location and the extent of this warming area within the wider Polar region, a substantial distinction can be made. The pronounced warming area covers less than 1/3rd of the Arctic area, but it extents well into the northern parts of the Greenland-, Norwegian-, and Barents Sea.

b) Greenland

It is widely acknowledged that Greenland went through a significant warming-up period. This is well demonstrated in the research work of R. Scherhag, in 1936, which indicates that temperature had increased with more than + 3°C from 1921-1930 (Scherhag, 1936). Our interest is to establish whether the brisk warming at Spitsbergen has any correlation with the observed warming in Greenland. The timing, development, intensity, and duration will be of big help for a better understanding of the Spitsbergen event. The main questions are:

- Where did the warming start at first, or did the warming start at several locations at the same time;
- Can a time gap be identified, and what time delay;
- How significant is the proportion of increase, and
- How long lasted a warming up period.

Although one can argue that each answer to any question needs to be evaluated and discussed in the light of the findings for Spitsbergen, we will discuss the matter in a comprehensive manner due to the fact that older material often raises only single aspects. It is furthermore necessary to be aware that Greenland's maritime climatology is not necessarily unison but differs between the East and the East Coast.

A start shall be made with the findings of one of the prominent names in meteorology Jacob Bjerknes (1897-1975), according to whom a warming in the East of Greenland occurred after 1920 (Bjerknes, 1959)[35]. The indicated period of time in question is from 1920 to 1930/32. Bjerknes assessed seawater temperature data in the North Atlantic as it follows:

- North of about 57° North the trend in sea temperature has been slightly upwards. Actually this change resulted from a brief but strong upward trend in the 1920s
- The warming of the waters in the far northern Atlantic was much more sudden and short range than farther south.
- Essentially, it lasted only from 1920 to 1930 in Greenland waters. (Bjerkness, 1959).

---

[34] The distances to Spitsbergen are roughly ca. 1200 km each.

[35] The Bjerknes paper indicates for the same time: "A somewhat similar brisk upward trend, stating as late as 1920, is found in the Labrador current on the Newfoundland Banks in close proximity to the main stem of the Gulf Current. asserted that the Labrador Current in the West of Greenland had shown a brisk upward trend, starting as late as 1920.

5. The warming event in details

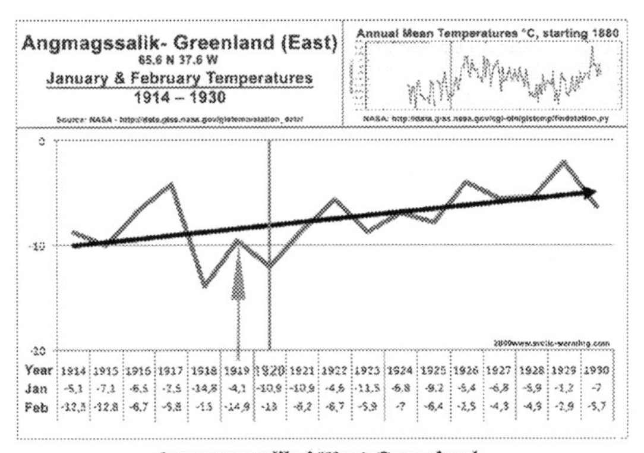

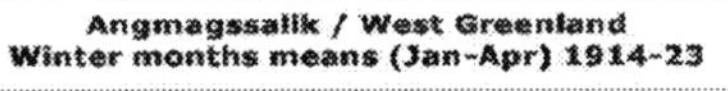

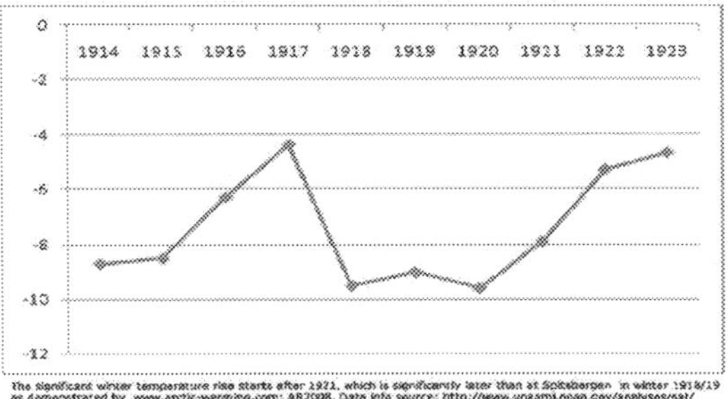

The significant winter temperature rise starts after 1921, which is significantly later than at Spitsbergen in winter 1918/19 as demonstrated by www.arctic-warming.com; AR2008.

Angmagssalik-East Greenland
Annual mean temperatures °C
Source: NASA-GISTEMP - http://data.giss.nasa.gov

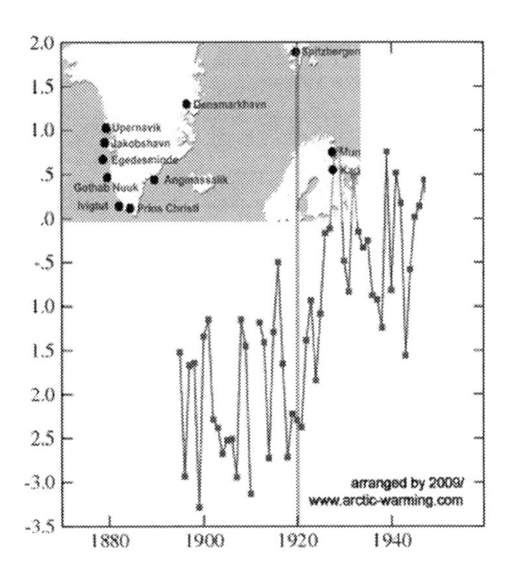

Thus the general Greenland picture indicates that a substantial warming took place after 1919, but that the warming period was limited to about one dozen years. The Greenland warming rate during the decade 1920-1930, shall have been 50% higher than during the 1995-2005 period. (Chylek, 2006)

For the East coast of Greenland data information are particularly sparse. Many observation stations were installed only in 1920. For example, at Myggbukta, 73°29'N, 21°34'W, data had been recorded since 1922; whereon the summer average temperature had shown a typical warming trend, which lasted until 1930/39 (Kirch, 1966), while generally winter data indicated a warming trend since 1923, which decreased since the period 1929/38 (Kirch, 1966). It was further reported that the last 'bad ice year' in the Greenland Sea was 1923 (Manley, 1944).

At least one station is available at the East Coast, Angmagssalik at the southeast of Greenland, for which Nasa-Gistemp provides a detailed record since 1880. The data confirm Kirch's analysis. The annual temperature rise did not start before 1922[36]. Actually, only the change from the cold years 1919-1921 is fairly pronounced with about 1,2° against the two years 1922/23. That can neither be regarded as big, nor sudden, but it is nevertheless noticeable. Any significance diminishes further if reviewed against the temperature data for the months January and February. For the time period 1918-1920 the annual amplitudes do hardly impress with significance. With regard to identifying a trend, it seems that one could identify the year 1920 as the starting point of a rising trend, which would compete with the annual trend within a range of 1-2 years. For timing the warming at the East Coast of Greenland it might also be of help to bring a graphic from the fishing sector. The cod fishing could only increase harvest after the fish population could grow in warmer water and that had been suddenly available after 1920 (Carruthers, 1941). The trend change in the early 1930s is remarkable as well.

[36] Annual mean temperatures at Angmagssalik (according Nasa-Gistemp): 1919 -2.22; 1920 -2.29; 1921 -2.38; 1922 -1.38; 1923 -0.93, 1924 -1.84; 1925 -1.08, 1926 -0.17; 1927 -0.12

2007

M. Stein (2007), Warming Period off Greenland during 1800-2005:

Their potential Influence on the Abundance of Cod and Haddock in Greenland Waters;

J.Northw.Atl.Fish.Sci.; Vol.39; p. 1-20

Extract from Abstract: (Nuuk- air temperature graphic added) Greenland and its adjacent waters are located at the northern boundary of the Sub-polar Gyre and thus subject to climatic variations within this gyre. It is suggested that periods characterized by regional shrinkage of warm water masses within the Gyre adversely affect the propagation of gadids from upstream Icelandic waters to Greenlandic waters, and periods of regional dilatation of warm water masses within the Gyre are favourable for developing gadid stocks in Greenlandic waters. Ocean temperatures off West Greenland show a significant upward trend, which is considerably higher than that for the North Atlantic Basin.

_____The paper states concerning: **Nuuk air temperature anomalies**; Very cold conditions were seen during the 1880s to 1910s., Warming began around 1920; and presently Nuuk enjoys warm conditions as during the 1926-30 period.

______The paper states concerning: **Nuuk sea surface temperatures** (see modified M.Stein graphic); the 1890s appeared as a decade with warm SSTs; the first decade of the new century showed shrinking of the Gyre, and only East Greenland waters were warmer than normal during those times; from 1915 onwards, warming increased; and a warm Gyre was seen in the second part of the 1920s.

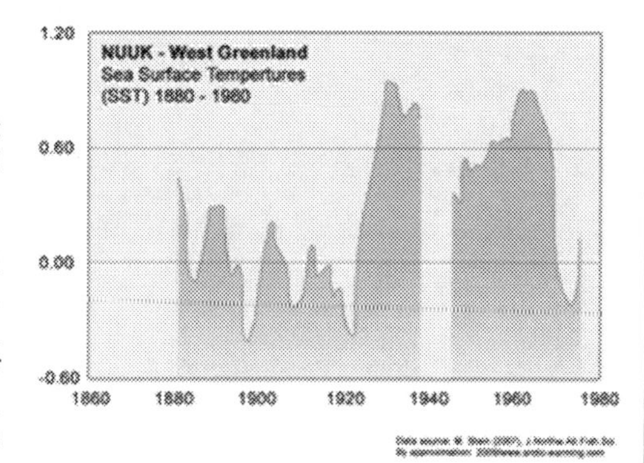

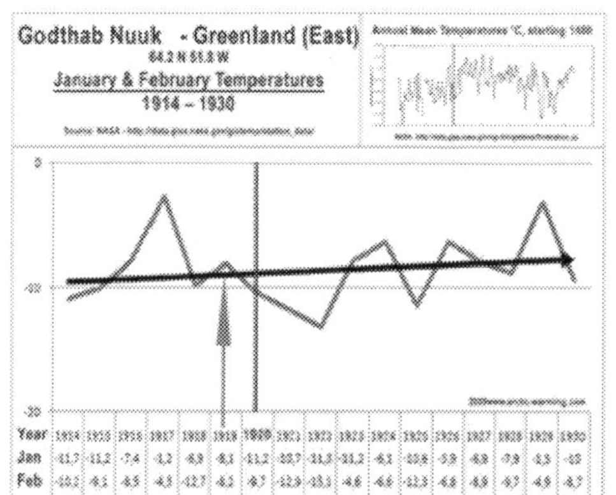

Godthab Nuuk-West Greenland.
Annual mean temperatures °C
Source: NASA-GISTEMP - http://data.giss.nasa.gov

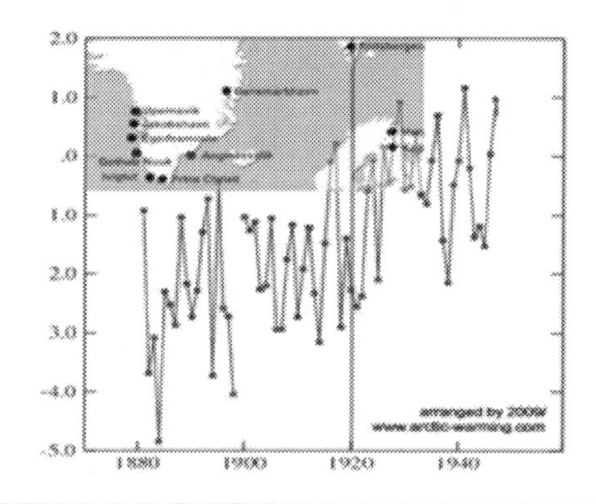

Comment:

____The series do not show an immediate correlation between the big warming at Spitsbergen during winter 1918/19, but a time lag of about two to three years. The SST increase at about 1922 is remarkable.

____ The first warming impulse seem to have been coming from the North, while only during the second part of the 1920s a warm Gyre had been observed.

5. The warming event in details

In the range of Greenland's West Coast at least four locations offer a longer dating record since the late 19th Century. On one hand the region is of greater interest because the West Greenland Current uses all the way along the West coast up to the northern part of the Baffin Sea before turning south toward Newfoundland. At Jakobshavn it is possible to make at least two observations. The January-February series (1914-1930) has no distinctive change close to 1920, while the annual temperature record does, but it actually became pronounced starting only in 1921/22, and lasting only until ca.1930. That the record does not show a more prominent result for the winter months, could be due to the extent of sea ice during winter time, which made the location more continental, with little or no effect on the air temperatures. A recent research asserted that although the last decade of 1995-2005 was relatively warm, almost all decades within 1915 to 1965 were even warmer at both the south-western (Godthab Nuuk) and the south-eastern (Angmagssalik) coasts of Greenland (Chylek, 2006). The temperatures at other stations can be discussed in the same manner as done in the case of Angmagssalik that would also lead to the same results. With some confidents one can say that the warming trend did not start before 1920, and did not last longer than the early 1930s.

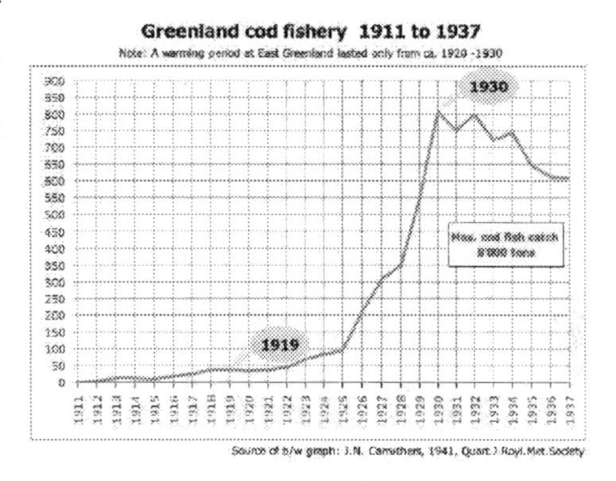

c) Sub-polar North Atlantic

Trend of Average Annual Sea Surface Temperature
From 1890/1897 to 1926/1933

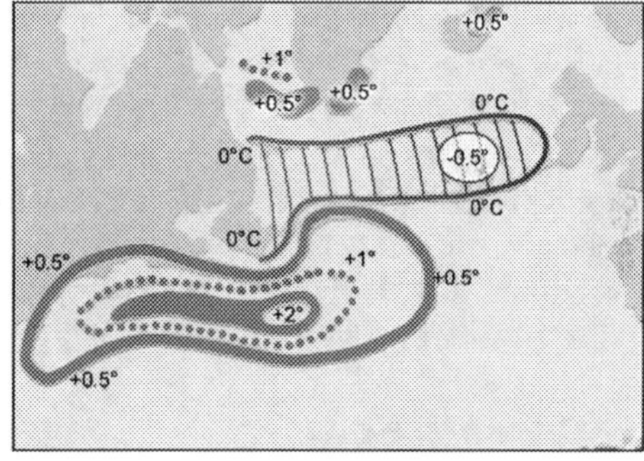

Image data source: J. Bjerknes (1959), ed Bert Bolin: The Atmosphere and the Sea in Motion. Here: 2009/www.arctic-warming.com

One way to temper Greenland by the ocean water system is the discussed East and West Greenland Current coming from the North and heavily influenced by very cold but less saline Arctic water. But what about any water coming from the South or Southeast, where the warm Gulf Current, and the Irminger Current, is not very far away. Had the warming of the Arctic been generated in the Gulf of Mexico and spread via the Atlantic Gulf Current to the Arctic? The region is described as sub-polar, and it is practically the only basin through which the heat from the South can be carried by the system of oceanic currents towards the pole because of the open boundary between the Atlantic and the Arctic oceans.

R. Scherhag addressed the question already in 1937, and analysed water temperatures in the North Atlantic, with the conclusion that the Gulf Current at source had been warmer during the years 1926-1933 than during the period 1912-1918, and that a similar increase had been observed in the English Channel, albeit smaller, for the time period 1920-1927. (Scherhag, 1937; and 1939):

Twenty years later J.Bjerkness picked up the issue but differentiated the situation. He said (extracts):

- "The warming of the Atlantic waters continues in a tongue extending eastwards from the edge of the Banks (Newfoundland) over the southern European coast.

- Between 50°N and 57°N the trend in the sea temperature has been slightly negative. That is, in fact, the only region within the whole Gulf Stream system where the long range trend of warming fails to show up.
- This belt of negative trend to temperature is traversed by the strongest, farthest left branch of the Gulf Stream system which is heading for Iceland.
- North of about 57°N the trend in sea temperature has been slightly upward. Actually that secular change results from a brief but strong upward trend in the nineteen-twenties which compensates the accumulated effect of a preceding long and slow downward trend." (Bjerknes, 1959). Note: The southern cape of Greenland, Kapp Farvel, is 60°N.

Bjerknes' analysis and conclusion show that the warming of the water around Greenland's coasts has not been coming and generating from the south, but north of about 57°N. In this respect a further conclusion should be noted:

- From the northern data collection can be seen that the sea surface temperatures near Greenland culminated in the early nine-teenthirties, while from Iceland to the British Isles the maximum water temperatures seem to have occurred in the early nine-teenforties. Despite the irregular downward trend, following the culmination, the general level of sea surface temperatures remains well above the low recorded in the northern areas around 1920.

Meanwhile it seems undisputed that off south west Iceland significant low SST temperatures had been observed in the early 1920 before a brisk temperatures rise occurred together with the air temperatures with a short delay only (Eythorsson, 1949), and that the Faroe were even later effected, since the early 1930s (Drinkwater, 2006).

In conclusion it seems obvious that the general situation in the middle North Atlantic range did contribute marginally to the warming of Greenland starting in 1920 and ending in the early 1930s. The one decade of Greenland warming has quite obviously been coming from the Northern North Atlantic, the Nordic Sea area. From where if not from the West Spitsbergen Current could a warm water stimulus had been coming from? The short period of one dozen years strongly indicates in this direction.

d) Barents Sea

Although the distance between the entrances to the Barents Sea in the South and in the North are only 1000 km apart, the two locations had had a quite different temperature experience in winter 1918/19 and the subsequent years of the 1920s. As an example may first serve the Russian record from Kandalsaka near Murmansk. The annual temperature did not show any change between the years 1918 and 1919, but only from 1919 to 1920. This time the increase is significant (+1.5°C) but neither exceptional within the record nor impressive concerning the following years which showed downwards. Concerning the core winter months (J/F), the difference by +2° from 1918 to 1919 is modest. Kandalaska did not experience an extraordinary temperature rise. The second example is Vardø, in the most north-eastern corner of Norway and close to the North Cape Current, which keeps the sea usually sea ice free[37]. Here the winter temperatures (D/J/F) actually increased between the years 1918 and 1919 (+2°C), and decreased slightly the following year, with a modest rise until 1930. The annual record is more interesting at least with

[37] See Chapter 7, Special Page; "2009 Question to Bengtsson"

Kandalaksa
Annual mean temperatures °C
Source: NASA-GISTEMP - http://data.giss.nasa.gov

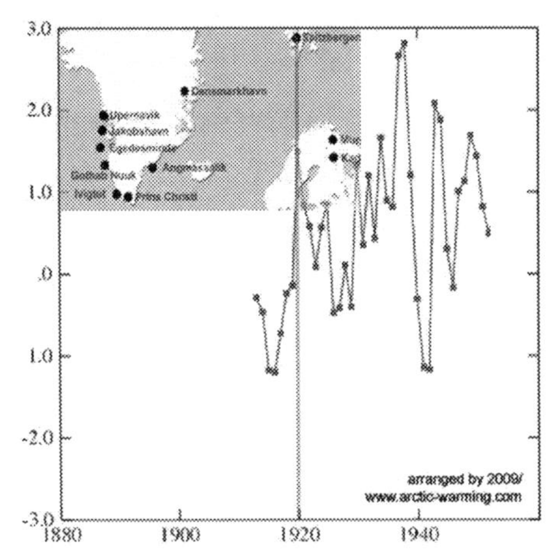

KARA SEA - AUGUST - SEA ICE
August ice extent anomalies
six-year running means - by 1000 km2

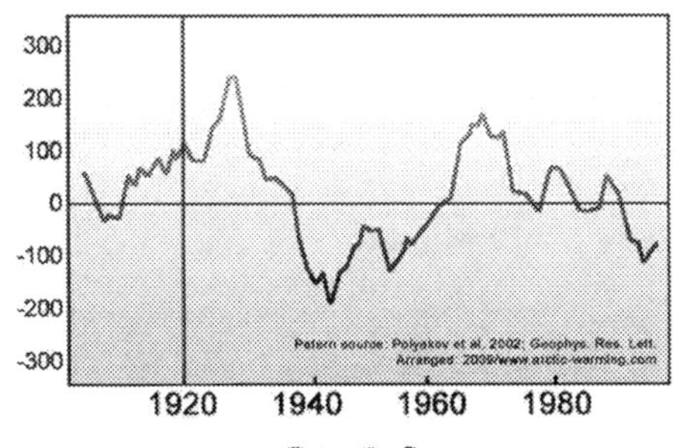

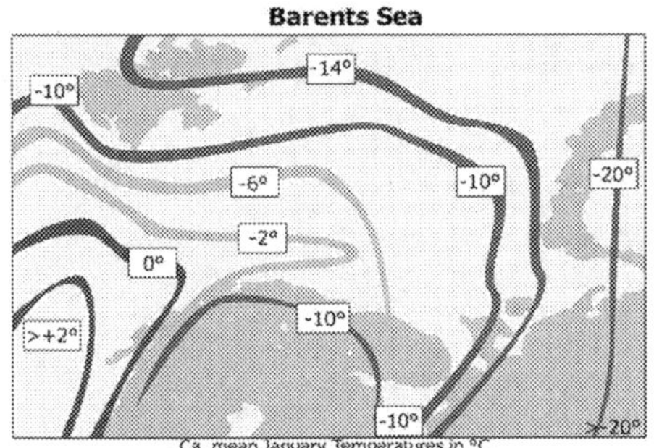

regard to one aspect, namely the rise between 1919 and 1920, which indicates a general shift. The mean temperature before winter 1918/19 and there after shows a shift from one degree. On the other hand the record shows that the Vardø location could not have contributed to the big warming of Spitsbergen in winter 1918/19. The fact that the region did not contribute to the sudden Arctic warming immediately does not necessarily mean that there was not a verifiable contribution over the total warming period until 1940. But one conclusion can already been drawn at this stage, that any detectable influence would be more modest rather than powerful or even decisive. This aspect will be raised again later.

In addition to the remarks in the previous chapter (relative shallowness, inflow of polar sea water, etc), the eastern part of the Barents Sea was usually fully ice covered during the core winter season 90 years ago. Sea ice is rare in the south-western portion of the Barents Sea. In April, however, the sea can be 75 percent ice covered. This extensive ice cover may play an important aspect when looking at special issues. As a matter of fact in the 1920s particularly high temperature had been observed at Franz-Joseph-Land, and at Novaya Zemlya, which require an explanation. However, in this region the increase was only observed since 1927 and 1928, only since than one could speak of "incontestable evidence of a progress warming" (Schokalsky, 1936) in this region.

As the open sea can release much more heat than an ice covered area, one simple explanation may be that the 'heat' had been coming with the flow of western wind from Spitsbergen, with its permanent supply with warm Atlantic water, which can transfer its heat to the air due to the sea ice free tongue up high in the North. The usually extent of sea ice here and elsewhere in the north-eastern part of the Barents Sea hardly allows any other explanation, at least for the winter period. Any claim that warm Atlantic water had "managed" to cross the Barents Sea, and remained warm enough to heat theses remote Russian islands through the existing sea ice, sounds little convincing and needs to be proved. The distant from Spitsbergen (West coast) to the West coast of Franz- Joseph-Land is ca. 800 km, respectively less than 2000 km to Novaya Zemlya. In particular, such locations as Vardø and Kandalaska (see above) would have felt if warm Atlantic water had entered the Barents Sea by mass and travelled to that very distant location. From this region a contribution to the initial warming period can be excluded, and it might be questionable if it is claimed that warming of the Barents Sea and Kara Sea consisted of 2° annually until the mid of the 1920s (Kelly, 1982). Only since 1920 temperature observation was taken at Boar Island, a location roughly half way between

2007

Pohjola, V. (2007);

„Arctic Warming – a Perspective from Svalbard",

Global Change News Letter, No. 69, p. 9 -12

The paper provides an indication that the annual average (Feb-Jan, 5-year series) at Spitsbergen increased from about 1918/19 to 1921/22 by four degree, but that the trend from the 1930s to the mid 1990s showed a general cooling, contrary to a warming over the whole century. With these facts at hand it seems strange when the paper says:

______With the IPCC report released during 2007 the scientific community is adding confidence to the relation between global warming and the boost of the greenhouse effect via anthropogenic emissions. Modelling work after the previous IPCC report have shown that the Arctic region is likely to warm up faster than the global average, and that the Arctic may be one of the regions to have the quickest response to global warming.

______The reason for this Arctic warming is debated, but was likely an effect of enhanced atmospheric circulation triggered by heat excess in its source region.

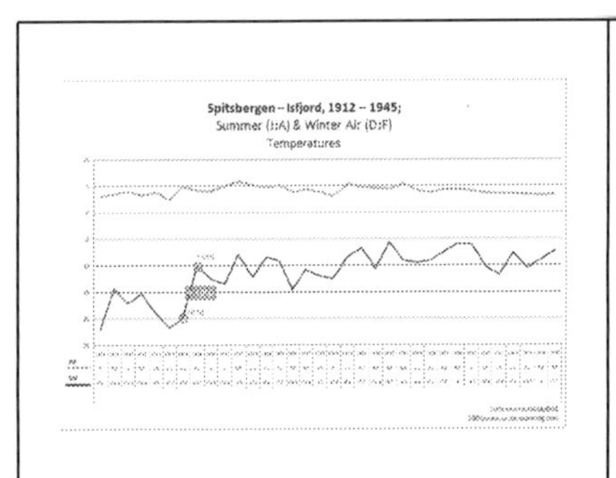

LEFT:
Air-temperature
Dec. to Feb.
1917/18=-21,7°C
1918/19=-10,1°C
RIGHT:
Annual mean

Pohjola:
The North Atlantic drift is a powerful contributor of heat into the Nordic Seas and further into the Arctic Ocean, where the northern branch of the drift splits at Svalbard.

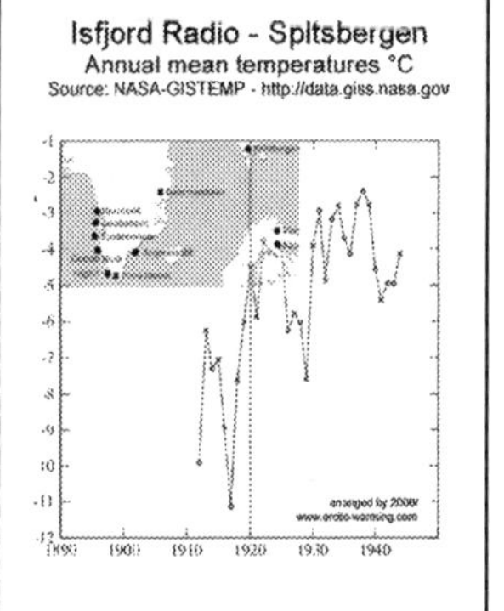

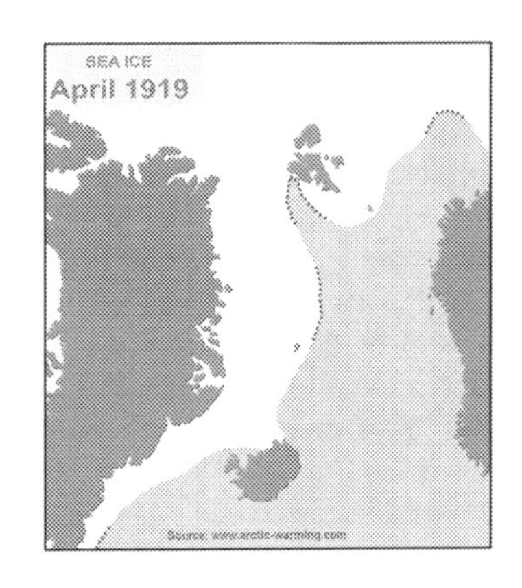

How can such a high rise in winter occur when the location is almost enclosed by sea ice?

Pohjola: One of the "hotspots" in this Arctic warming may be Svalbard, due to the fact that this archipelago is positioned right where the Arctic front separates the polar and extra-tropical air and water masses.

Pohjola: How well the Svalbard region senses global climatic variability is exemplified by the much larger warming the archipelago experienced in the 1920–30s event than what was recorded at other North Atlantic/ Arctic sites.

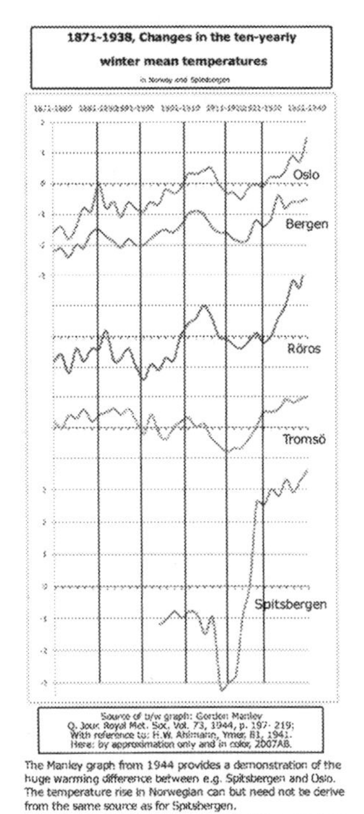

The Manley graph from 1944 provides a demonstration of the huge warming difference between e.g. Spitsbergen and Oslo. The temperature rise in Norwegian can but need not be derive from the same source as for Spitsbergen.

www.arctic-warming.com

The strong warming of North Europe – 1936 to 1938

the North Cape and Spitsbergen. The recorded winters of 1920 and 1921 (January/February) were significantly colder than the following ones, except for the winter of 1928/29, until the winters of 1939/40 and 1940/41. In so far, the warming of Boar Island started with the winter of 1921/22. The annual means are fairly equal (1° to 3.2°C, between 1920 and 1932), getting warmer only after 1932 until 1939. (Kirch, 1966). Although it is a small temperature increase over a two decade period, the Barents Sea did not contribute in any significant way to the Arctic warming, and at the time of commencement, not at all, respectively only modestly over the subsequent two decade period.

e) Europe

In Europe the temperatures increased only very slowly but steadily since 1920. Although the 'natural climatic system' of continental Europe had definitely not been the source of the Arctic warming during the 1920s and 1930s, a brief introduction will indicate that not necessarily all warming has been coming from the North. However, here we concentrate merely on the correlation between the Arctic warming and warming in Europe since winter 1918/19, while considering any correlation between the Spitsbergen warming with Europe's treatment of its coastal waters due to naval activities during WWI from 1914 to 1918 is discussed later.

A convincing piece of evidence is the data set for Norway, demonstrating that the difference between Spitsbergen and the city of Tromsö, only about 800 km away, is tremendous (Manley, 1944). The shown graphic indicates the slowly advance of rising temperatures at Bergen and Oslo. This could be interpreted as a clear indication that not only one but at least two forces had been at work. This is indirectly confirmed by the observation that the NAO (North Atlantic Oscillation) has weakened during the 1920s and remained weak for the whole period of the warm Arctic anomaly (Bengtsson, 2006). It is obvious that Europe, neither the land nor the sea areas had been in the driving seat, due to the fact that the warming came in creeping pace only. The start of Europe's warming since 1920 had presumably not been noticed in practice by anyone. Even nowadays it seems impossible to establish that there had been any signs available in the late 1910s that an increased warming could be starting soon. However, it happened, but definitely later as at Spitsbergen, and simultaneously in time with Greenland.

It lasted until the winter of 1939/40. The warming was so pronounced that autumn 1938 was the warmest, together with 1772, 2000 and 2006, in the last 500 years (Xoplaki, 2006). Summer temperatures also had risen substantially; with 1°C. Autumnal temperature rises in the 1930s were local and observed in Scandinavia and western part of maritime Russia only (Polyakov, 2003). No other continental Northern Hemisphere region experienced a similar rising trend. The United States data records, which had a modest warming until 1933, saw a decrease in temperatures since then. The correlation with Greenland is perfect. In Europe the warming continued for another six years. The air temperatures over North Europe from 1936-1938 indicate that higher temperatures were observed in the Baltic Sea region (Scherhag, 1939).

Although it can be assumed that at the end of the 1930s Europe experienced the highest annual temperatures in several hundred years, it can be questioned whether there had been an exclusive connection between the Arctic warming and the warming in Northern Europe. Iceland got warmer since the mid 1920s (Eythorsson, 1949) and the Faroe in the northwest of Scotland since the early 1930s (Bjerknes, 1956). Therefore, the middle Atlantic seems to have played no significant role in the warming of Europe. Although the warming of Northern Europe between the two World Wars is evident, the main warming impulse came from the North.

E. Spitsbergen as a Heating Spot

If one asks whether the heating-up spot is to be found at Spitsbergen, we would answer: yes. The information supplied sustain this affirmative answer. Nothing demonstrates this better than the data record taken at Spitsbergen since 1912. If one reviews the January/February temperature difference between the winters of 1913/14 and of 1919/20 (ca. + 15°C), or from the winters of 1916-1917 to the winters of 1919-1920 (ca. + 22°C), the results are not only extraordinary, but they reveal that the "shift" took place in 1918, respectively in the winter of 1918/19 (Hesselberg, 1956). This is emphasized by the comparison between the data recorded from 1912, until WWI ended in November 1918 (ca. – 4.3°C), and thereafter (ca. +3.8°C), including the winter of 1925/26.

In the summer of 1918 the seawater temperatures had already reached unusual values: +7°C to +8°C at the West coast of Spitsbergen (Weikmann, 1942). During the winter of 1918/19 the temperatures varied considerably. There were long periods in November and December 1918 with temperatures close to zero degrees, 4 days with temperatures above zero in November and 7 days in December[38]. In January 1919, the temperatures did not reach –5°C for 14 days, and five days were frost-free. The annual mean (1912-1926) with minus 7.7°C suddenly jumped to an annual average of minus 5.4°C in 1919, representing a plus of 2.3 degree. The corresponding figures provide for January a difference + 8.6°C, which indicates that the sea was able to transfer a lot of heat into the air. However, during February–April 1919, the temperatures were well below the average (ca. -6°C), with a large ice-cover far out into the sea. But that did not affect the significant warming that had started a few months earlier.

F. Summary

The interpretation of the reference material indicates that an outstanding warming-up phenomenon can be located with precision at Spitsbergen, and the exact timing is within a range of a few months, with the core month January 1919. Such a precise date cannot be found for other location. As there was no simultaneous temperature jump during the corresponding time period elsewhere, it is possible to assert with certainty that Spitsbergen represents the first place where the Arctic warming started at the beginning of the 20th century. Therefore the next chapters will subsequently examine: WHY did the "greatest yet known sudden temperature rise on earth" (Birkeland, 1930) presumably occur. After all, the "severe warming" at Spitsbergen in 1918 did not come from "nowhere", nor did the subsequent 'climatic change'. The event needs to be explained, either that the sudden change was due to internal dynamics of the sea body, or due to an external force, or a combination of both.

[38] Average temperatures (1912-1926) for December minus 14.4°C: in December 1918 minus 7°C.

Chapter 6
How is the agitation in the Arctic explained?

A. Can a 'climatic revolution' be analysed?

a) Are the IPCC Reports of any help?

To investigate the early Arctic warming that occurred one hundred years ago, one would presumably get not far if taking a core conclusion of the IPCC serious: *"The Earth's global mean climate is determined by incoming energy from the Sun and by the properties of the Earth and its atmosphere, namely the reflection, absorption and emission of energy within the atmosphere and at the surface"*(IPCC, 2007). Followed from this understanding that IPCC had only little to say about the Arctic "climatic revolution", namely that the *"Average Arctic temperatures increased at almost twice the global average rate in the past 100 years. Arctic temperatures have high decadal variability and a warm period was also observed from 1925 to 1945" (IPCC, 2007).* With such statement the IPCC demonstrates how superficially one of the most striking climatic shifts of the 20th century is handled. Due to the fact that the sun is not and cannot be the direct source of the extraordinary warming during the Arctic winter, this leads inevitable to the prime source of heat supply, the ocean and the seas. With more attention to this very principle polar region condition the IPCC assessments would provide a more reasonable explanation, and presumably lead to the understanding, that "Climate is the continuation of the oceans by other means"[39], whereby 'means' denote: heat and water supply from the ocean and seas to the atmosphere.

b) Is 'Chaos' a hindrance?

One can hear it often: The atmosphere is a chaotic system and as such it is inherently unpredictable. As weather is forced by the heat energy of the sun, the atmosphere is, therefore, unstable and non-linear. These two characteristics are the crucial components of chaos (Palmer 1991). Discussing the chaos of weather in modern science is about climate change projections. Can computer models tell us anything about the mechanism of climate in the future? Definitely not, as long as models cannot handle the past. As indicated by the output from IPCC in the previous paragraph, one of the most decisive climatic events during the last 150 years computer models seems to have not explained anything. Is the work with weather data of little promise to detect the reasons for the early Arctic warming during the winter season? No, the ocean is the dominating source of warming the Arctic during the time the sun is not shining in the Polar region.

One can furthermore hear: Does the chaotic nature of oceanic circulation limit the predictability of climate, just as the chaotic nature of atmospheric circulation is known to limit the predictability of weather?" (Covey, 1991). Indeed, circulation in the oceans and seas is extremely complex and based on many physical characteristics, namely on temperature, salt concentration and density. These conditions create a permanent forcing within the water body of one square-centimetre, a square-kilometre, a regional

[39] The climate definition matter is discussed in detail at: http://www.whatisclimate.com/

2007

World Climate Report introduced on the 16th of October 2007 the issue:

Greenland Climate: Now vs. Then, Part I. Temperatures

http://www.worldclimatereport.com/index.php/2007/10/16/greenland-climate-now-vs-then-part-itemperatures/

Subject of a Hot Topic at: www.arctic-warming.com, on 18th Oct.2007

This site refers to the early arctic warming at the start of the last Century, which actually was a Spitsbergen warming commencing suddenly in winter 1918/19. Now World Climate Report is willing to demonstrate that Greenland was as warm, or warmer, than it is presently, wondering that this fact seems largely ignored by alarmist scientists. That is good news and may be also of significant assistance to the efforts of this site.

Particularly useful are the given references of Greenland temperature data. The most interesting are from a location at Greenland's East coast named Angmagssallik, which has - according NASA – an air temperature set since 1895. This might help to identify clearly where and when the extreme warming started in the Northern North Atlantic. In Part C, Section: The warming event in detail , this site concluded that the warming commenced in 1918, latest in January 1919.

The reproduces winter temperature-set for Angmagssallik, and the corresponding two graphs (for winter and annual mean around the year 1920) show clearly that the warming at East Greenland started one or two year later, as the winter/summer temperatures at Spitsbergen. Attention should be also given to the two graphs showing the minimum and maximum sea ice cover, which usually made Angmagssalik an inland location up to 400 kilometres away from the open sea towards the end of the winter season.

During winter the remote archipelagos Spitsbergen is for short and long-term weather making and changing a unique place. Due to the warm water from the Gulf Current a small section remains sea ice free, and that is the reason why the early warming started in winter and started here. Hopefully World Climate Report continues vigorously by elaborating the warming of Greenland, but is it also able and willing to look across the Greenland Sea to Spitsbergen, considering why it all started there in winter 1918/19.

Annual Mean Temperatures
in Celsius before 1920 and after 1920

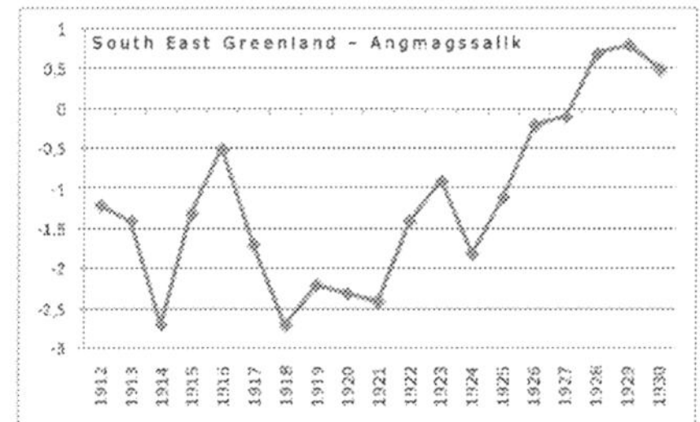

Around the year 1920 the air temperatures at Angmagssalik (ca. 65° N, 37° W) are fairly steady and only start rising briskly since ca. 1922, almost 2-3 year later than at Spitsbergen in winter 1918/19. AB2008
Data source: http://www.unaami.noaa.gov:

Angmagssalik / East Greenland				
	Jan, Feb March April	Four months	+ Dec.//Year	5 months
1914	-5.1,-12.3,-10.6, -6.8	-34.8/-8,7	-9.3,//1913	-8,8
1915	-7.1,-12.8,-10.0, -4.2,	-34.1/-8,5	-7.3,//1914	-8,3
1916	-6.5, -6.7, -6.5, -5.5,	-25.2/-6.3	-4.7,//1915	-6,0
1917	-2.5, -5.8, -5.5, -3.6,	-17.4/-4,4	,-10.3,//1916	-5,5
1918	-14.8,-13.0, -5.1, -5.1,	-38.0/-9.5	,-13.7,//1917	-10,3
1919	-4.1, -14.9, -8.5, -8.6,	-36.1/-9.0	-6.7,//1918	-8,6
1920	-10.9,-13.0, -8.7, -5.7,	-38.3/-9-6	-6.2,//1919	-8,9
1921	-10.9, -6.2, -7.5, -6.8,	-31.4/-7.9	-8.9,//1920	-8,0
1922	-4.6, -6.7, -5.9, -3.9,	-21.1/-5,3	-6.7,//1921	-5,6
1923	-11.5, -5.9, -1.3, 0.0,	-18.7/-4,7	-3.9,//1922	-4,5

Data info from: http://www.unaami.noaa.gov/analyses/sat/

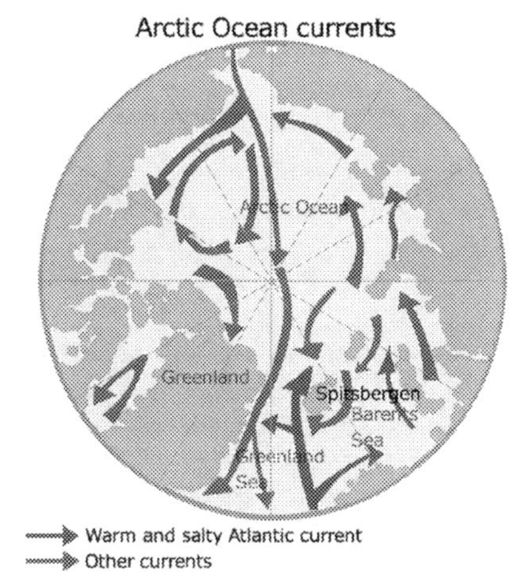

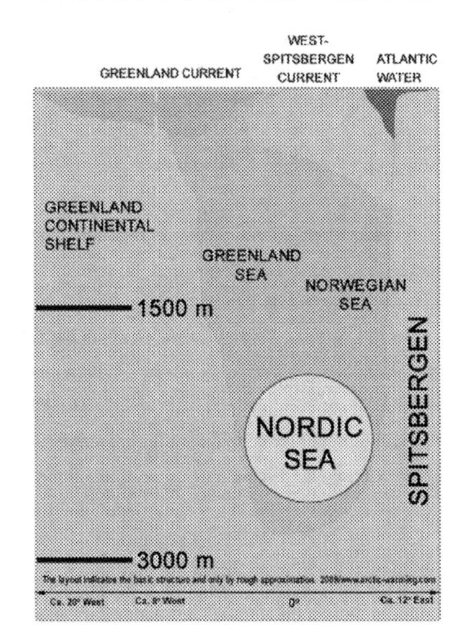

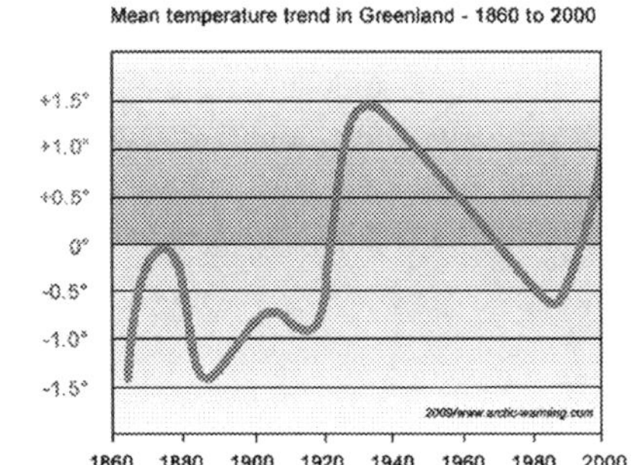

sea, or the entire global oceans. However, this does not rectify to assume the internal oceanic processes as chaotic in nature. Due to its physical structure, a huge cold water body below a very, very thin warmer sea surface, the oceans are extreme stable over time and space. The mean temperature of less than four degrees Celsius has presumable not changed more than one or two degrees over many 100'000 years.

Over millions of years the ocean structure has ensured that the global mean temperatures have not deviated more than about 10° Celsius from today means. Without the oceans and seas the daily global temperature would cover a range of 300° Celsius. The obvious inability of modelling the ocean processes does not mean that analyzing the event at Spitsbergen is hampered by a chaotic oceanic behaviour. If the modellers had accepted that weather modelling will only be promising if based on a comprehensive set of permanently collected ocean data, they could have explained the warming of the Arctic 90 years ago since long. They should have realized that the term "climate" is not well define as the average weather, but that "climate" should be understood as an immediate result of released heat and water vapour from the oceans and seas, every second, and over all time periods. The best region on earth for such climatic studies is the polar region, and particular if such studies can be based on a unique event as the warming in winter 1918/19.

c) The Arctic, Al Gore and better understanding

Al Gore said in July 2007: "Just in the last few months, new studies have shown that the north pole ice cap – which helps the planet cool itself – is melting nearly three times faster than the most pessimistic computer models predicted"[40]. A few months later, he, together with the IPCC, were honoured with the Peace Nobel Prize in December 2007 for "for their efforts to build up and disseminate greater knowledge about man-made climate change, and to lay the foundations for the measures that are needed to counteract such change".

The awarding Committee was presumably not aware that neither of the two Nobel laureate had hardly anything achieved in explaining why the Arctic warmed the Northern Hemisphere during the first half of the last century. Al Gore had paid the North Pole a visit by submarine before becoming Vice President of the U.S.A: "We were crashing through that ice, surfacing, and I was standing in an eerily beautiful snowscape, windswept, and sparkling white, with the horizon defined by little hummocks, or 'pressure ridges' of ice that are pushed up like tiny mountains ranges when separate sheets collide. But here too, CO2 levels

[40] The New York Times (2007); Al Gore, "Moving Beyond Kyoto", Sunday, the 1st of July 2007, WK 13.

are rising just as rapidly, ...As the polar air warms, the ice here will thin; and since the polar cap plays such a crucial role in the world's weather system, the consequences of a thinning cap could be disastrous" (Al Gore, 1992).

Al Gore's North Pole visit took place about 1990. As almost two decades have passed, one can only wonder that he and his fellow alerter on global and arctic warming still understand so little about at least the early part of the warming in the Arctic. Instead he claimed recently:

- "We – the human species – have arrived at a moment of decision."
- "What is at risk of being destroyed is not our planet itself, but the conditions that have made it hospitable for human beings".
- "We – all of us – now face a universal threat. Though it is not from outside this world, it is nevertheless cosmic in scale." (NYT, 1 July 2007).

How can someone talk about a cosmic scale threat if he is not able and obviously not interested in understanding a 'climatic revolution' that occurred under the eyes of modern science only 90 years ago? The explanation is presumably simple. Whoever is talking about climate without thinking comprehensively in the terms conditioned by the oceans and seas, will get not far in understanding what makes climate tick and where and when human activities are changing climate due to altering the 'natural structure and course' of the oceans and seas.

d) Any analysis is to be based on ocean impact

The question, whether the early Arctic warming can be analysed, intended to stress the ocean issue. No doubt, the impression that the oceans play an important role in weather and climate matters is widely spread, particularly among the average laymen. But when it comes to thinking in oceanic matters comprehensively, science is handling the questions fragmented and in a piecemeal manner. In this way one quickly misses the point. For example back in 1991 a paper asserted that "changes in the intensity of the thermocline circulation, and hence its poleward heat transport, would have a significant effect on global climate" (Weaver, 1991).

Who if not this heat transport is the most likely source of the "climatic revolution" in the Arctic before the 1940s. But neither IPCC, nor the scientists working with models and chaos theories, and also not Al Gore, seem to be able to consider the warming of the Arctic from an oceanic perspective.

Even leading Arctic research experts seem to have to go some way to put more focus on the sea areas surrounding the North Pole. They will have difficulties to explain the present situation as long as they cannot explain what had happened in the past. The following section provides a selection of recent findings concerning the warming 90 years ago, with the emphasis on the 'presentation' of the early warming, whether the ocean issue is addressed, and which explanations are given. In the subsequent chapter the main aspect of the early Arctic warming and its causation will be discussed.

B. Recent findings in Arctic research

The following excerpts are selective and do not necessarily present only individual results of the authors but may include quoted conclusion. However, the selection intents to provide a principle picture on whether the authors are aware of the big warming at Spitsbergen since winter 1918/19, respectively the Arctic warming, and how it is presented. For subsequent work the interested reader is advised to consult the original texts.

Overpeck, et al., 1997

From 1840 to the mid-20th century, the Arctic warmed to the highest temperatures in four centuries. This warming ended the Little Ice Age in the Arctic and has caused retreats of glaciers, melting of permafrost and sea ice, and alteration of terrestrial and lake ecosystems. Although warming, particularly after 1920, was likely caused by increases in atmospheric trace gases, the initiation of the warming in the mid-19th century suggests that increased solar irradiance, decreased volcanic activity, and feedbacks internal to the climate system played roles.

Jones, et al., 1999

The Northern Hemisphere warmed between about 1880 and 1940, and cooled after 1940.

The global surface air temperature (SAT) has increased with 0.6 ° C since 1861, with a slightly higher rate of warming in the twentieth century.

Kelly, et al., 1999

Trends in Arctic temperatures have been broadly similar to those for the Northern Hemisphere during the study period. The Arctic variations were, however, greater in magnitude and more rapid.

The overall range of the long-term variations in the temperature of the Arctic has been greater in winter.

In the winter record, the 1910s rather than the 1880s was the coldest decade. 1917 and 1918 were notably cold years especially in the winter records. The 1910s was the coldest decade in winter.

The annual data show a drop of 0.45°C from the previous decade. During the final years of this decade, warming began in the Barents Sea and Kara Sea regions.

The 1920s was a transitional decade with strong warming affecting most regions. The Barents and Kara Seas had warmed by about 2°C annually.

Polyakov, et al., 2003

The higher temperatures in the Arctic, during the 1930s–40s and the recent decades, and the lower temperatures, during the 1960s–70s and prior to the 1920s may be associated, at least partly, with the positive and negative phases of the LFO (low-frequency oscillation).

The seasonal differences in the alternate warming periods (1920s–30s and from the 1980s to the present) and cooling period (1940s–50s) are striking.

6. How is the agitation in the Arctic explained?

The warming during the 1920s–30s was very fast in spring, autumn and winter, but much weaker and slower in the summer. The period between 1918 and 1922 displays exceptionally rapid winter warming.

The rapid autumnal temperature rise, in the 1930s, was a local phenomenon that was observed only in Scandinavia and in the western part of maritime Russia.

This variability appears to originate in the North Atlantic and is likely induced by slow changes in oceanic thermohaline circulation.

However, SAT records demonstrate multi-decadal variability, which is stronger in the polar region than at lower latitudes. This may suggest that the origin of this variability may lie in the complex interactions between the Arctic and North Atlantic.

The Arctic temperature was higher in the 1930s–40s than during the recent decades, and hence a trend calculated for the period beginning with 1920 and going up to the present actually shows a cooling trend.

The complicated nature of Arctic temperature and pressure variations determines the difficulty of understanding the causes of the variability, and of evaluating the anthropogenic warming effect.

Polyakov, et al., (2004)

The Arctic surface air temperatures were highest in the 1930s – early 1940s.

The normalized AWCT (Atlantic Water Core Temperature) shows warm periods from late 1920s to 1950s.

Bengtsson, et al., 2004

The huge warming of the Arctic, which started in the early 1920s and lasted for almost two decades, is one of the most spectacular climate events of the 20th century.

The Arctic warming from 1920-1940 is one of the most puzzling climatic anomalies of the 20th century.

During the peak period of 1930-1940, the annually averaged temperature anomaly from the area 60°N-90°N amounted to around 1.7°C.

Whether this event is an example of an internal climate mode or of an externally induced phenomenon (by enhanced solar effects) is presently under debate.

Here we suggest that natural variability is the most likely cause, reduced sea ice-cover being crucial for the warming.

This warming was associated and presumably initiated by a major increase in the westerly and south-westerly wind, north of Norway, this leading to an enhanced atmospheric and ocean heat transport from the warm North Atlantic Current, through the passage between northern Norway and Spitsbergen, into the Barents Sea.

6. How is the agitation in the Arctic explained?

We suggest that the warm Arctic event just happened by chance, through an aggregation of several consecutive winters with pronounced, high latitude westerly in the Atlantic sector.

It is interesting to note that the increasing high latitude at which westerly usually flow wasn't related, this time, to the North Atlantic Oscillation, which was simultaneously weakening.

What consequences may have the findings of this study for the possible evolution of the Arctic climate? Notwithstanding an expected overall climate warming, it is suggested that the Arctic climate would be exposed to considerable internal variations during several years, initiated by stochastic variations of the high latitude atmospheric circulation and subsequently enhanced and maintained by sea ice feedback.

Johannessen, et al., 2004

SAT (surface air temperature) observations and model simulations indicate that the nature of the arctic warming in the last two decades is distinct from the early twentieth-century warm period. It is strongly suggested that the earlier warming was natural internal climate-system variability, whereas the recent SAT changes are a response to anthropogenic forcing.

Two characteristic warming events stand out, the first from the mid 1920s to about 1940 and the second starting in 1980 and still ongoing.

The early twentieth-century warming was largely confined to north of 60N.

Both the 1920–39 and 1980–99 warming phenomena are more pronounced during winter for the high Arctic.

The anthropogenic forcing in the 1920s–1930s was by far too weak to generate the observed warming – the change.

The GHG (green-house-gas) forcing in the early decades of the twentieth century was only 20% of the present.

We theorize that the Arctic warming in the 1920s/1930s and the subsequent cooling until about 1970 are due to natural fluctuations internal to the climate system.

Serreze, 2006

Since the mid-1800s, global average surface air temperature has risen approximately 0.7°C. It is widely believed that at least part of this warming arises from increased concentrations of infrared absorbing gases in the atmosphere, the so-called greenhouse effect.

Substantial high-latitude warming from about 1920 to 1940 was followed by cooling until about 1970.

A trend calculated from 1920 to present, however, yields a small Arctic cooling. Over the period 1901 to 1997, the difference in the Northern Hemisphere trend and that for the Arctic is statistically insignificant.

6. How is the agitation in the Arctic explained?

It nevertheless seems clear that the warming in recent decades, the warming in the earlier part of the 20th century, as well as the cooling between them are more pronounced in high latitudes than the Northern Hemisphere as a whole.

The earlier warming has no summer signalled at all, and exhibits a strong latitude dependency. In summary, we conclude that:

> (1) In sharp contrast to the high-latitude warming in the earlier part of the 20th century, the recent warming is part of a global signal, suggestive of external forcing;
>
> (2) Arctic amplification of this SAT signal, as well the observed decline in the sea ice cover, has been strongly influenced by low-frequency climate variability, especially that associated with the NAM and PDO[41];
>
> (3) The NAM and PDO cannot neatly explain all of the changes.

Overland, 2006

What is the comparison of recent decades with earlier warm periods in the Arctic such as the 1920s –1940s.

It should be noted that the Arctic north of 66°N represents a small fraction of the globe and that it lies at the northern limit of major storm tracks; thus Arctic is subject to considerable seasonal to decadal atmospheric variability or 'climate noise'.

The Arctic also includes major feedback processes. The most important is albedo feedback where loss of snow or ice increases the absorption of solar radiation by land or ocean. A second one is cloud – radioactive feedback where increase open water creates increased moisture flux to the atmosphere, this creating more clouds and a shift in radiation balance.

The Arctic is also influenced by external forcing from volcanoes, changes in solar activity, and anthropogenic sources.

At present all three factors can be considered 'causes' of Arctic climate: natural variability, internal feedbacks, and external forcing, but with unknown relative importance and interaction.

The earlier warming shows large region-to-region, month-to-month, and year-to-year variability suggesting that these composite temperature anomalies are due primarily to natural variability in weather systems.

Overland, 2008

A meridional pattern was also seen in the late 1930s with anomalous winter (DJFM) SAT, at Spitsbergen, of greater than +4°C. Both periods suggest natural atmospheric advective contributions to the hot spots with regional loss of sea ice.

---

[41] Northern Annular Mode (NAM) and Pacific Decadal Oscillation (PDO).

2009

„A Focus on Climate During the Past 100 Years"

by S. Brönnimann et al. (ed); 2008
Brönnimann; Luterbacher, J.; Ewen, T.; Diaz, H.F.; Stolarski, R.S.; Neu, U. (Eds.) 2009, XVI, p. 364;

A comment to the editor's "Introductory Paper" at http://www.1ocean-1climate.com, 27. Dec. 2009 (Extract; graphs & reference not shown)

The 'Introductory Paper' is fair in acknowledging that there are many open questions, but has little proof to offer if it claims that the past 100 years are significant for "the changeover of a climate system dominated by natural forcing to a climate system dominated by anthropogenic influences". The paper presents the matter as fact which is elsewhere formulated as question: "Was the 1910–1945 trend a result of "natural variability" and the 1950–2003 trend an "anthropogenic" warming?" (p.9). The paper says also that the introduction gives an overview of the book in the context of recent research, highlighting some of the key findings and concepts (p.2), but rarely does. Instead a mingle-mangle of other findings is presented with little, if any, critical review.

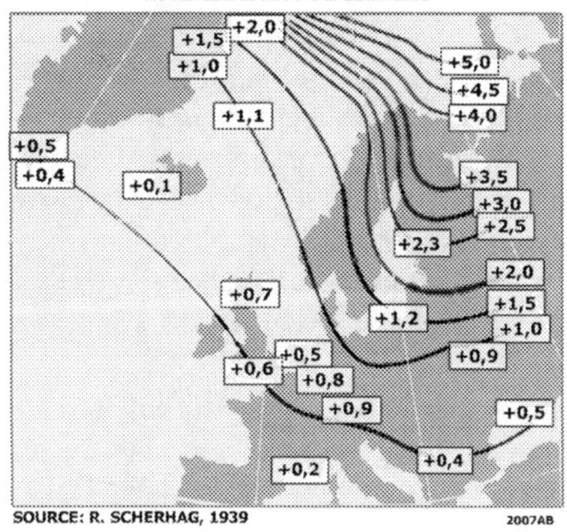

For example:

INTRODUCTION p.2: During the past 100 years the Arctic experienced two pronounced warming periods. Between 1915 and 1945, annual mean temperatures increased by about 1.8°C. This period was followed by a cooling and a more recent warming, which started around 1970 and is still ongoing.

COMMENT A:

___Any assessment of the recent climate should acknowledge that a global warming trend had started at the end of the Little Ice Age around 1850, because the two pronounced warming periods mentioned could be part of one warming-up since 1850, interrupted by a three decade cooling phase (1940-1970). By global warfare?

___The sudden increase of temperatures started in winter 1918/19 and not 1915.

___The warming was initially not global, but had an impact in the USA until about 1933.

___The warming did not last until 1945 but ended with three extreme Northern Europe winters immediately after World War II commenced.

INTRODUCTION p.8/9: The 1910–1945 (warming) trend was most pronounced over the Atlantic and North America, while Europe experienced a winter cooling. In contrast, the 1950–2003 period exhibited a strong winter warming of the northern hemispheric land masses. What may have caused this difference in climatic response?

COMMENT B:

___The here reproduced graphic for the time period Nov.1936 to Oct.1938 indicates the trend was quite different as claimed by Brönnimann.

___The period between 1940 and 1970 was strongly influenced by a global cooling.

___And is there a difference in climatic response? Yes & No!

___YES: The early warming had been caused by the West Spitsbergen Current, while the second warming was actually marking the end of the interrupted warming trend from 1940 to 1970s presumably partly due to the naval warfare during WWII.

___NO: Because the discussed changes had been generated and controlled by the oceans and seas.

C. The issues debated by Arctic science

a) The ignorance of older research

Did the scientific disciplines of meteorology and climate come in being only few decades ago? The work of modern scientists in the field of Arctic research seems to confirm the question. None of the mentioned papers has analysed the considerable work of their forefathers. Their long list of used references rarely uses research work older than 10 years.

Prominent names in meteorology from the 1930s, like Birkeland, Brooks, Scherhag, Schokalsky, Manley, Carruthers, etc. are seldom mentioned, and if so, only randomly. None of their interesting work is analysed, evaluated, or discussed. The extraordinary temperature rise at Spitsbergen, the date, the extent, the duration at this station and in the wider region have neither been emphasised nor elaborated. The extremely important distinction of summer and winter temperature in this region was never made to a core issue for research work, nor is the extensive sea ice cover from Kap Farwell/South Greenland, Iceland, and Bear Island in late spring made a discussion point.

It seems that modern science believes that the world of Arctic climatology had to be newly created somewhat 30 to 40 years ago. But even if they thought that their discipline would need to be newly invented, it would nevertheless have been necessary to prove the case by demonstrating that the research work of the 1930s and 1940s had been superficial, wrong, incompetent, and useless. But simply ignoring the work is too little to do serious research on understanding how the Arctic climate is ticking.

But even more fundamental aspects were ignored, respectively not seen. One of the most astonishing examples is the indifference of many research work shown when it comes to the marked difference between summer and winter temperatures. Although a number of papers acknowledge that there has been a marked higher temperature winter level versus the summer level, this issue was never picked up for thorough investigation. Although arctic winter warming cannot be directly enhanced by a solar effect, Overland and others do not question nor even hesitate to bring solar activities in play.

The superficiality is topped by a complete ignorance of seasonal air temperatures of Arctic coastal or inland stations in relation to the status of sea ice relevant at that location. At most general freezing and melting aspects are mentioned, but the extreme weather and climate relevant seasonal icing in the Northern North Atlantic and Barents Sea has not been elaborated in any of the papers mentioned above. One paper, for example, merely mentions that "reduced sea ice-cover being crucial for warming". No wonder that the researchers missed to elaborate and to explain the special situation of Spitsbergen throughout the 150 years since the end of the Little Ice Age, but particularly the early warming of 90 years ago, and the subsequent warming since the 1980s.

The astonishment is enhanced by the fact that some researches have no problem to acknowledge that the warming since the 1920s:

- was greater in magnitude and more rapid, than the recent one;
- displays exceptional rapid winter warming;
- the seasonal differences are striking;

- has no summer signal at all;
- is one of the most spectacular climate events;
- is one of the most puzzling climatic anomalies,

but do not rest until finding an answer to each of the raised points. Based on those initial general remarks, some further 'explanation' given in the referred paper will be picked up and discussed.

b) About the suddenness of the early event

None of the papers recognises or investigates the most remarkable feature of the early warming, namely the suddenness. The suddenness, which we could pin-point as having happened –statistically- in January to February 1919, is the source of information that ticker question and cry for answers. At a remote archipelagos as Spitsbergen a sudden warming during winter would inevitably leave only few options for the identification of the causation. It seems impossible to come up with something else than the ocean body. But even that would need to be qualified more precisely, as a sea body like the Norwegian Sea itself would presumably never be able to produce such a big warming suddenly and to sustain it over two decades. The warm Atlantic water travelling toward the North Pole with the Spitsbergen Current would have quickly caught attention. The research of Arctic warming could commence.

For example: One researcher asserts that "this variability may lie in complex interactions between Arctic and North Atlantic". But the logic would have told one that the more interaction had happen, the less likely it would have allowed happening with suddenness.

Another example is certainly the frequent reference to "natural variability". One can not base a research merrily on longer period statistics without evaluating the statistics on lower frequencies as well. Any detailed analysis of the rise of winter temperatures at Spitsbergen shows that the jump was not natural, but "unnatural".

c) Circulation variations – Brooks Question 1938

One paper suggested that the Arctic climate would be exposed to considerable internal variations during several years, initiated by stochastic variations of the high latitude atmospheric circulation and subsequently enhanced and maintained by sea ice feedback. No word on what caused the alleged circulation variations. No word on what essential conclusion can be drawn. A stochastic process is one wherein the system incorporates an element of randomness as opposed to a deterministic system.

Other researcher said that the temperature variations might be due to: "at least partly, with the positive and negative phases of the low-frequency oscillation (LFO)". LFO means to vary between alternating extremes, usually within a definable period of time. Such notion explains not the suddenness, nor the magnitude, and neither the long duration over two decades. The notion does not lead to the location Spitsbergen, does not put the emphasis on Northern North Atlantic as the determining factor, and miss the point that this warming could have only be initiated and sustained by the West Spitsbergen Current.

The problem is not new but haunted already the researchers in the 1930s. What had caused the warming of the Arctic? R. Scherhag, presumably one of the most keen and competent researcher on the issue, was one of the few who came up with at least one assertion, which the American meteorologist C.E.P. Brooks, immediately questioned. In 1937 R. Scherhag wrote:

> *"The greater mildness of winters observable in the temperate zone during the last hundred years, accompanied by an increase in atmospheric circulation, has, during the last fifteen years, led to an extraordinary rise in temperature in the arctic regions, which in its turn has been accompanied by a corresponding retreat if the ice and a higher temperature in the sea. (Scherhag, 1937)*

In 1938 Brooks asked the right question, which neither Scherhag, nor modern science has answered yet:

> *"Attributing the recent period of warm winters to an increase in the strength of the atmospheric circulation only pushes the problem one stage further back, for we should still have to account for the change of circulation." (Brooks, 1938)*

Indeed! Since long such question concerning circulation changes and variation should not have been placed as presumptions, as mentioned in the previous paragraphs, but answered.

Brooks question is also not answered if, for example, one paper names instead the changed circulation a major increase in the westerly and south-westerly wind, north of Norway, and thereon coming up with the assumption that this lead to "an enhanced atmospheric and ocean heat transport from the warm North Atlantic Current, through the passage between northern Norway and Spitsbergen, into the Barents Sea" (Bengtsson, 2004). Presumably, the authors want to say what also Brooks expressed in 1938, when he wanted to explain his statement on accounting for a change of circulation:

> *"Moreover, it is almost equally plausible to regard the change of circulation as a result of the warming of the Arctic, for open ice conditions in the Arctic Ocean are favourable to the formation of depressions. More probably the increased circulation is both cause and effect of the warmed Arctic; high temperature causes storminess and decrease of pressure in high latitudes, which in turn is associated with stronger wet winds in middle latitudes, driving an excess of warm sea water into the Arctic and raising the temperature still further."(Brooks, 1938)*

Although Brooks was already on the right way, he seems himself aware that this is not the needed explanation on what caused the circulation to change in the first place. The cited recent paper has hardly more to offer as Brooks before 70 years. At least Brooks conclusions is presented in general terms, while the recent paper name the Barents Sea without any elaboration whether this fairly shallow sea, with an average depth of 230m, can initiate a very sudden warming, and sustain a warming of a longer period of time. This issue will be discussed more deeply in the next chapter.

d) The non sensual use of "natural variability"

Almost all papers relate the early warming in the 1920s partly or primarily to:

- natural variability in the weather system;

2007

Andrew C. Revkin, The New York Times, on 2nd of October 2007,

"Arctic Melt Unnerves the Experts"

http://www.nytimes.com/2007/10/02/science/earth/02arct.html

The stark shrinking of the Arctic ice cap during summer 2007 was the article's concern, expressing –inter alias – the following:

___Over all, the floating ice dwindling to an extent unparallel in a century or more, by several estimates.

___One geologist summarized it in this way: "Our stock in trade seems be going away."

___Scientists are also unnerved by the summer's implications for the future, and their ability to predict it.

___Still, many of those scientists said they were becoming convinced that the system is heading towards a new, more watery state, and that human-caused global warming is playing a significant role.

___Other important factors were warm winds flowing from Serbia around a high-pressure system parked over the ocean.

A HOT TOPIC at www.arctic-warming.com,
on 26th October 2007 commented as it follows:

Recently the NYT journalist Andrew C. Revkin has elaborated the rapid decline of Arctic sea ice and found that the "Arctic Melt Unnerves the Experts". That needn't to be if more attention had been given to the extreme Arctic warming phase for two decades during the first half of the last Century.

Only few days ago WCR discussed the "Greenland Climate: Now vs. Then, Part I. Temperatures" (Subject to a 'Special Page' at this Chapter 6), before World War II because the island had been warm, presumably even warmer, than it is presently, wondering "that this fact seems largely ignored by alarmist scientists". The article demonstrates that within a few years in the early 1920s, the typical average temperature rose by about 2ºC. This is an important finding, but "peanuts" in comparison to the warming of Spitsbergen.

The even more important question may be in which region the warming actually started and when. On one hand one needs warm water that the Golf Currents supply to the West coast of Spitsbergen, on the other hand one needs to take into account the prevailing sea-ice conditions from December to April, as shown in a graphic. Actually, the East coast of Greenland is largely cut off from the open sea during the winter season, while in the West of Spitsbergen the sea remains ice free high into the North. The brisk warming trend only started after the year 1920, while the warming at Spitsbergen to the year 1918, latest to Januarys 1919.

By now one can only hope that the early Arctic warming receives further attention, as the climate debate should be based on understanding why the Arctic climate changed suddenly only 90 years ago. It is not enough just to claim that it happened in due natural course, as WCR did on the 22nd of October 22 2007 when discussing Andrew C. Revkin article.

6. How is the agitation in the Arctic explained?

- atmospheric variability or "climate noise";
- natural fluctuations internal to the climate system;
- considerable internal variations;
- feedbacks internal to the climate system.

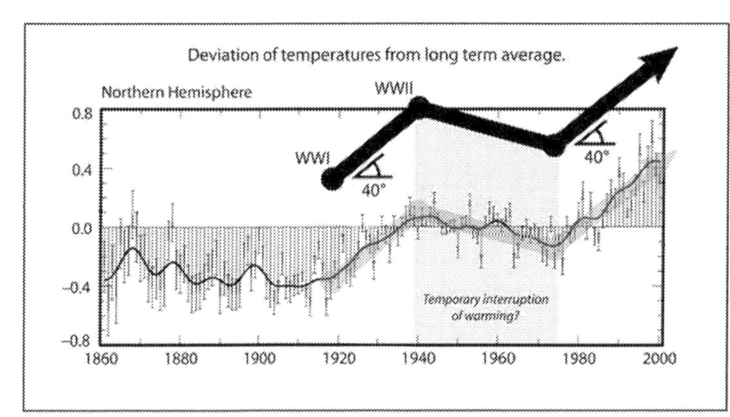

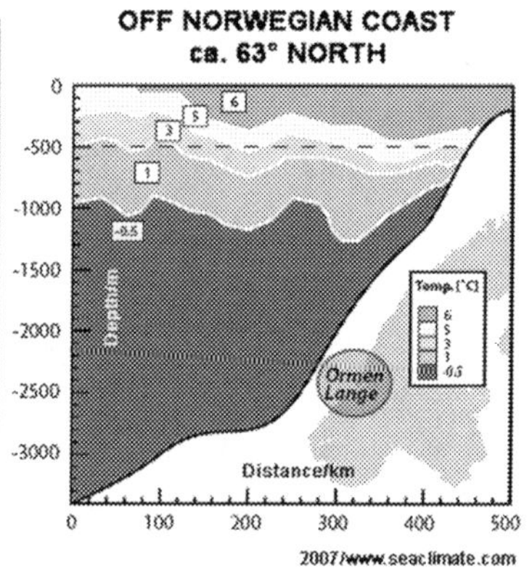

The sensitivity of the seas south of Spitsbergen during winter is indicated by the sea surface conditions

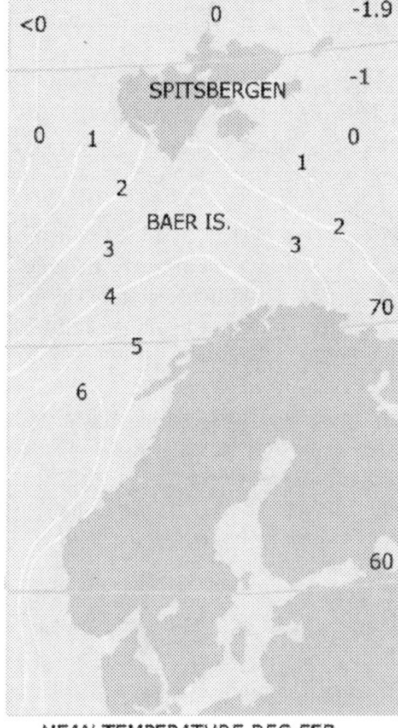

MEAN TEMPERATURE DEC-FEB

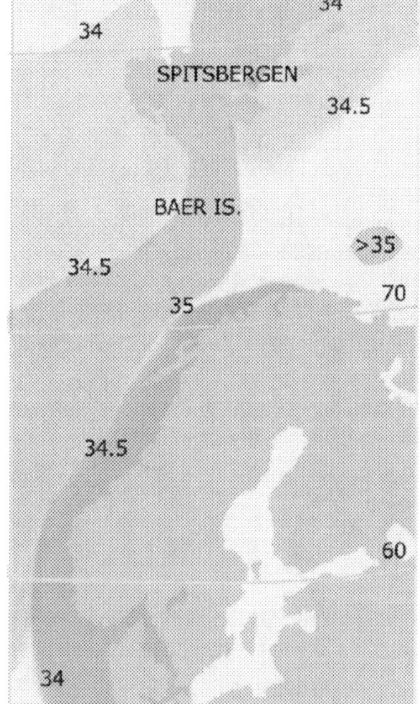

MEAN SALINITY DEC-FEB

These references explain nothing. Nowadays WMO[42] still defines climate as the average weather, while the global convention, on this matters has not a definition of climate at all[43]. Instead the "climate system"

[42] World Meteorology Organisation

is defined as: "the totality of the atmosphere, hydrosphere, biosphere and geosphere and their interactions". This does not paraphrase more than "global nature". Raising the matter here shall indicate the use of such term, or equivalent circumscriptions, does in no way contribute to explain anything of the situation in the Arctic since the late 1910s. The researchers in the 1930s seem to have had a more solid approach in mind, as none of them – as far as it could be observed – has ever assumed that this warming is due to "natural variability"|. More information on the term "climate" is available at a specialized web-site[44].

D. Summary

The question was, whether modern science is able to provide reasonable explanations on the reasons for the Arctic warming. This is definitely not the case for the warming that occurred 90 years ago. The assessments and conclusions remain superficial and hypothetic. The current arctic warming can presumably only reasonably be understood and explained by a thorough understanding of the warm and cold periods of the last century. It surprises that the sudden warming of the Arctic, although widely acknowledged, has never been seen as the presumably best study field, under the most promising circumstances, to reach evident results. Instead the matter is side stepped and made scientifically irrelevant by claiming that there is nothing to investigate due to the fact that it happed as a natural fluctuation. The next chapter shall show that this is not a sustainable approach, but that it is possible to prove that the causation can be named.

---

[43] United Nations Framework Convention on Climate Change, 1992; but defines instead in Article 1 the following terms:

1. "Adverse effects of climate change" means changes in the physical environment or biota resulting from climate change which have significant deleterious effects on the composition, resilience or productivity of natural and managed ecosystems or on the operation of socio-economic systems or on human health and welfare.
2. "Climate change" means a change of climate which is attributed directly or indirectly to human activity that alters the composition of the global atmosphere and which is in addition to natural climate variability observed over comparable time periods.
3. "Climate system" means the totality of the atmosphere, hydrosphere, biosphere and geosphere and their interactions.

[44] http://www.whatisclimate.com/

2008

Catherine Brahic on:

Arctic currents may be warming the world

In: New Scientist -26 April 2008, Magazine issue 2653, p. 12
http://www.newscientist.com/article/mg19826533.900-arctic-currents-may-be-warming-the-world.html

The article by Catherine Brahic demonstrates that there is substantial awareness of an oceanic impact on global warming. Not only ocean currents but Arctic currents may be warming the world. That sounds interesting, but only on first view. Some excerpts form the article will show that it is not at all meant as understanding the status of the oceans as the blue-print of climate. Nowhere else as in the polar region can warming by the ocean be better studied, when during the winter season the sun is not contributing. But the article explains its topic as follows:

__On top of the effect of human-made carbon emissions, natural changes in the warm ocean currents traveling to the icy north may be helping to heat up the entire northern hemisphere.

REMARK: Science says: Sun ray can easily pass through the outer atmosphere in order to reach the Earth, but much of the radiation escapes the atmosphere depends on the concentration of greenhouse gases (including carbon dioxide, methane etc) present.

QUESTION: What effect has CO2 at Spitsbergen during the winter season?

__Climate models produced by the Intergovernmental Panel on Climate Change (IPCC), which are tuned to reproduce the human-made greenhouse effect, predict the region should have warmed by 1.4 °C between 1960 and 2000. In fact, the Arctic's average air temperature rose by 2.2 °C.

REMARK: Not one word about the strong Arctic warming from 1919-1939

QUESTION: Can climate models work when they do not recognize nor include the warming at Spitsbergen?

__Vladimir Semenov of the Obukhov Institute of Atmospheric Physics in Moscow, Russia, says that ocean currents carrying warm water from lower latitudes into polar regions could have played a part in this increase. He analysed air temperature data from the north Atlantic, which revealed a cyclic pattern of highs and lows over the past century. He argues the length of such cycles must be explained by ocean currents, which also fluctuate over a timescale of decades.

REMARK: It is astonishing, that the transport of 'warm water into the polar region' is actually presented as if it is a new finding.

QUESTION: Is the suddenness of the warming at Spitsbergen since winter 1918/19 evidently not supporting the claim of 'cyclical pattern of currents'?

___ "It's an interesting idea - and the first time I've come across it," says Peter Wadhams, head of the Polar Ocean Physics Group at the University of Cambridge. Although the research is speculative, it is a plausible process, Wadhams says.

REMARK: The comment demonstrates where polar research stand, and how important it would be for the understanding of the climate change issue what had happened 90 years ago at the remote archipelagos of Spitsbergen.

QUESTION: Have Vladimir Semenov and Peter Wadhams ever read what Schokalsky, Brooks, Scherhag and others have said in the 1930s?

CONCLUDING REMARK: The title of Catherine Brahic's article sounds so promising, but the explanation hardly demonstrate that the mechanism in the Arctic is understood. It wouldn't have happen with more attention to the main features of the Arctic warming for two decades since the late 1910s. That is certainly not her fault.

Chapter 7
Where did the early Arctic Warming originate?

A. Suggested forcing mechanism

a) An introduction with a lecture given in 1935

On the 30th January 1935, the Royal Scottish Geographical Society honoured the President of the Geographical Society of the Soviet Union, Jules Schokalsky, with the Society's Research Medal. In his address he informed the Society that records provide incontestable evidence of a progressive warming of the Arctic Ocean:

> *"The branch of the North Atlantic Current which enters it by way of the edge of the continental shelf round Spitsbergen has evidently been increasing in volume, and has introduced a body of warm water so great, that the surface layer of cold water which was 200 metres thick in Nansen's time, has now been reduced to less than 100 metres in thickness." (Schokalsky, 1936)*

For this investigation it is now time to ask what might have forced the change in the polar realm. In previous chapters the location and time-period for the sudden Arctic warming 90 years ago has been established, which leads to the question what has or may have triggered the event. Neither Johannessen (Johannessen, 2004)., who recently assumed that the warming in the early part of the 20th century was probably a natural phenomenon, nor Bengtsson (Bengtsson, 2004), who asserted that this climatic anomaly was probably a result of the influx of warmer water into the Barents Sea (see below), can be of much help. Closer to the core issue came Polyakov (Polyakov, 2003), with the conclusion:

- *This variability appears to originate in the North Atlantic and is likely to be induced by slow changes in the oceanic thermohaline circulation.*
- *However, SAT records demonstrate stronger multi-decadal variability in the polar region than at lower latitudes.*
- *This may suggest that the origin of the variability may lie in the complex interactions between the Arctic and the North Atlantic.*

Although all three-research papers come up with a 'conclusion', none of them realises that the results are of little help. As already mentioned in the previous chapter, C.E.P. Brooks (Brooks, 1938) had expressed his disagreement with regard to R. Scherhag's assert on increased atmospheric circulation, as this pushes the problem one stage back because one should still have to account for the change in circulation. Brooks made a thoroughly right diagnosis. All the conclusions which have been previously quoted can be quickly questioned today using Books' comment that he made 70 years ago. What is different from Scherhag's suggestion is the fact that at least two of the quoted opinions make reference to the role that the sea might have played in the warming phenomenon.

7. Where did the early Arctic Warming originate?

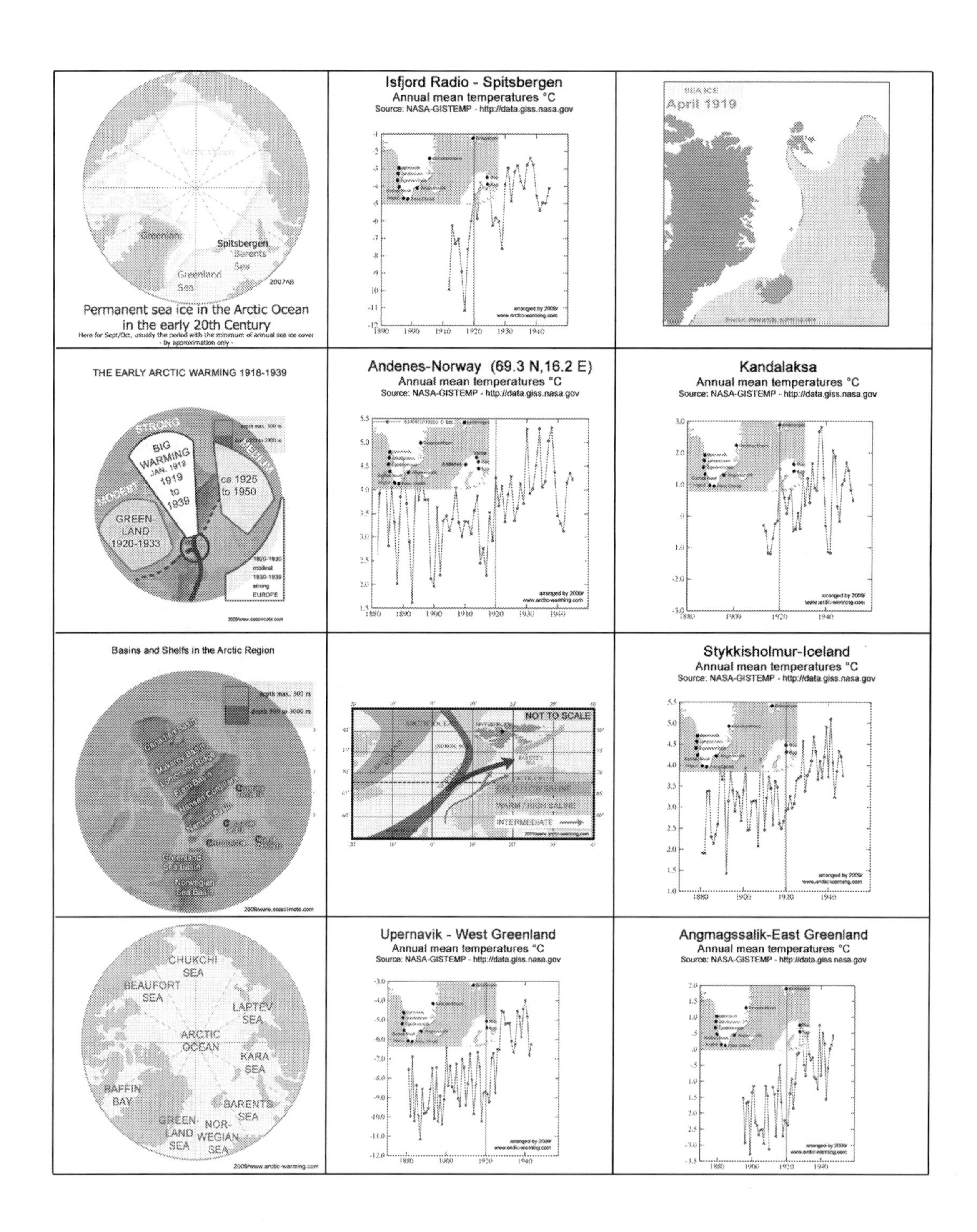

1982

Aagaard, Knut, (1982),

„The Climate Environment of the Arctic Ocean"

in: Louis Rey, The Arctic Ocean, Comite Arctique International, London, p. 69-81.

Extracts:

____Even the deepest portion of the Arctic Ocean is therefore not isolated from surface events in the seas to the south. The details of the deep lateral exchange are completely conjectural, however: Is the flow large or small? Does it occur in broad bands or slender ribbons? Is it confined to a thins near-bottom sheet, or is it a thick layer? Is it highly variable in time, or is it relative constant?

____We do not know very much about the cause of variability (salt and heat), nor about its consequences.

____So far we have not said anything about what happens to the Atlantic water within the Arctic Ocean. This situation is rather like a black box, for which we may know something about the input function but have neglected the response function. Until we determine the internal circuitry of the box, we shall not be able to make any useful analyses and predictions.

____The consequences of such onshore fluxes of salt and heat are not clear, but what is clear is the events in the north Atlantic can be transmitted years later to the shore of the far Pacific side of the Arctic Ocean.

____It may well be, for example that in the long run the Eskimos in Alaska will care considerably what the European puts into the Irish Sea.

____It is important to point out that we know next to nothing about the deep circulation of the Arctic Ocean.

____A rather direct Atlantic influence is therefore pervasive in the Polar basin, from the shelves to the abyss.

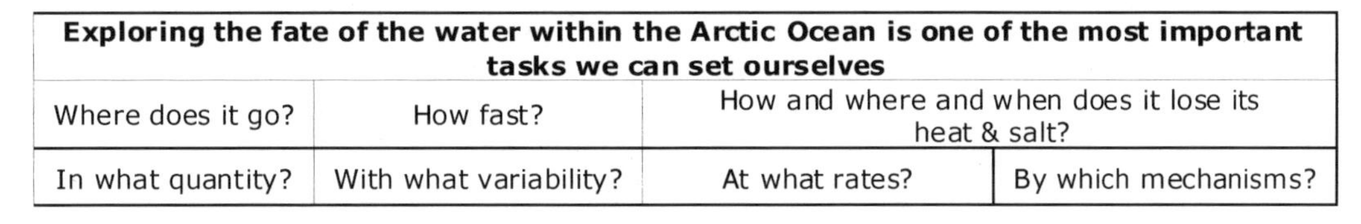

Exploring the fate of the water within the Arctic Ocean is one of the most important tasks we can set ourselves			
Where does it go?	How fast?	How and where and when does it lose its heat & salt?	
In what quantity?	With what variability?	At what rates?	By which mechanisms?

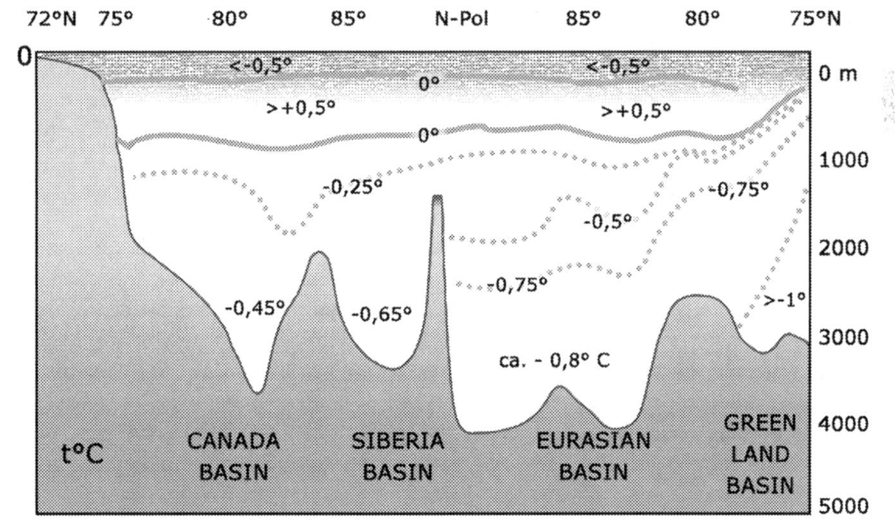

Source: Knut Aagaard (1982)

In many respects, the oceanography of the Arctic Ocean is at the point it was in the Atlantic 60 years ago, prior to the great modern oceanographic expeditions.

b) Slow change in ocean circulation – Oceans interaction

It seems that Polyakov et al. (Polyakov, 2003) have missed the identification of the most interesting points, especially when assuming that the variability might have been induced "by slow changes in oceanic thermohaline circulation". This notion neglects completely the fact that there must have been a very sudden and dramatic change in the oceanic interior. The previous analysis demonstrates this fact beyond any doubt. Obviously, the 'big warming' from the winter of 1918/19 could have been caused only by an extremely rapid change, so quickly that it has never been observed since weather and ocean records have begun to be registered, 200 years ago. The statement sustaining that the "variability appears to originate in the North Atlantic" is not very enlightening either, although the location where the "variability appeared" can be very precisely identified as being the island of Spitsbergen.

It is also difficult to agree with the affirmation sustaining that the *"variability may lie in the complex interactions between the Arctic and the North Atlantic"*. The problem derives particularly from the word "interactions" because the overriding relation between the two oceans is the one-way transport of warm water to the Arctic basin. The West Spitsbergen Current transports warm Atlantic waters to the North, through the Fram Strait into the Arctic Ocean, and, in the opposite direction, the East Greenland Current transports very cold low saline water and sea ice southwards. The features of the two currents are so different that one can consider them, in the widest sense, as very separate entities. While they run in opposite directions, there is inevitable and considerable mixing and interactions going on. But these mixing and interactions cannot be held responsible for the generation of such an extraordinary warming up event. Actually, the higher any interaction at the time period in question, the less significant would have been the warming up of Spitsbergen.

Increased interactions between different currents of the North Atlantic can be excluded. The most prominent currents flowing from South to North (off Norway's coast) and from North to South (off Greenland's coast) are separated by a distance of about 100 to 300 km. There is no claim sustaining that any significant warming, or cooling, or any other relevant change in weather statistics had been observed in the North Atlantic, along the Polar Circle (66° 34' N) or in the south of it, in the winter of 1918/19.

c) The wind induced Arctic warming

As it is still asserted nowadays that the early warming "was associated and presumably initiated by a major increase in the westerly and south-westerly wind, north of Norway, this leading to an enhanced atmospheric and ocean heat transport from the warm North Atlantic Current, through the passage between northern Norway and Spitsbergen, into the Barents Sea" (Bengtsson, 2004), this should raise astonishment. Such a statement needs to be challenged for a number of reasons, primarily for ignoring a principle rule already mentioned: T*he energy that maintains the atmospheric circulation is to a great extent supplied by the ocean* (Sverdrup, 1942). Any list of questions should certainly include such as:

- Where did the wind come from to push warm Atlantic water north-eastwards?
- Has the wind over the Norwegian Sea "access" to the warm North Atlantic water?
- How much water could be pushed by wind into the Barents Sea, respectively how much wind must be available, to increase the flow of water from West to East throughout the Barents Sea?

2004

Questions to Bengtsson et al. paper 2004

Concerning: Enhanced wind driven oceanic inflow into the Barents Sea

Bengtsson, Lennart (2004), Vladimir A. Semenov, Ola M. Johannessen, The Early Twentieth - Century Warming in the Arctic—A Possible Mechanism, Journal of Climate, page 4045-4057:

The authors assert that the temperature rise in the Arctic in the early 1920s was caused by enhanced wind driven oceanic inflow into the Barents Sea. The Abstract of the paper summarizes this aspect – inter alias – as follows:

1. The huge warming of the Arctic that started in the early 1920s and lasted for almost two decades.
2. By analyzing similar climate anomalies in the model as occurred in the early 20th century, it was found that the simulated temperature increase in the Arctic was caused by enhanced wind driven oceanic inflow into the Barents Sea with an associated sea ice retreat.
3. Observational data suggest a similar series of events during the early 20th century Arctic warming including increase westerly winds between Spitsbergen and the northernmost Norwegian coast, reduced sea ice and enhanced cyclonic circulation over the Barents Sea.
4. It is interesting to note that the increasing high latitude westerly flow at this time was unrelated to the North Atlantic Oscillation, which at the same time was weakening.

The author's conclusion should be questioned, at least on reasons as follows:

- Was the jump of winter temperature at Spitsbergen from 1919-1923 not by far to high and sudden for having been caused by oceanic inflow to the Barents Sea.
- Is the Barents Sea with a mean depth of 230m not to shallow to 'receive' a substantial amount of warm Atlantic water. See figure on water temperature at 30m depth in August (right – below).
- How can it be explained that the temperature at Spitsbergen exploded, while the rise in Norway was moderate over the years until 1939.
- If the Barents Sea had indeed received a stronger inflow, should the temperature record at Vardø not be substantial higher? (see the Vardø-graph)
- Do the data from Vardø not clearly indicate that there had been a modest system shift around 1919 (see the Vardø-graph, annual mean, indicating the different levels) of about one degree?

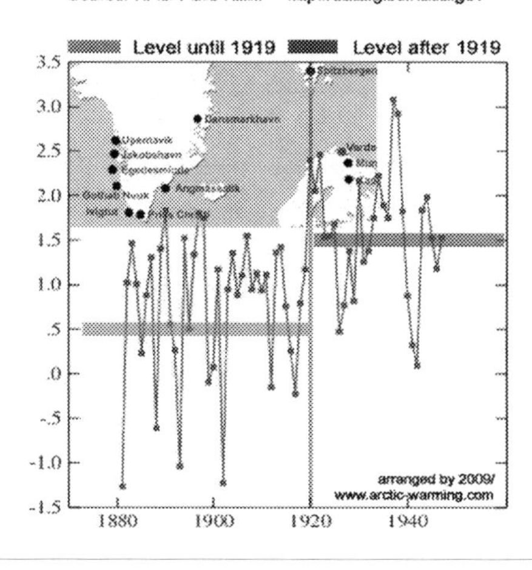

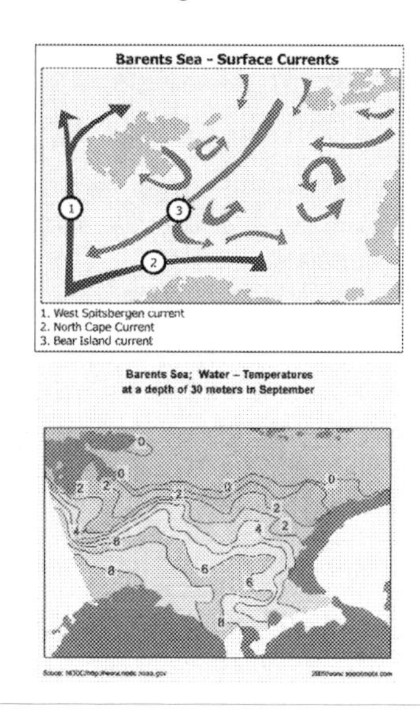

7. Where did the early Arctic Warming originate?

- How can wind influence the flow in the Barents Sea over the time period in question when the sea is at least partly covered by sea ice?
- If the wind pushed water into the Barents Sea between Norway and Spitsbergen, should there not a simultaneous high rise in temperature at the North Cape as there was at Spitsbergen?

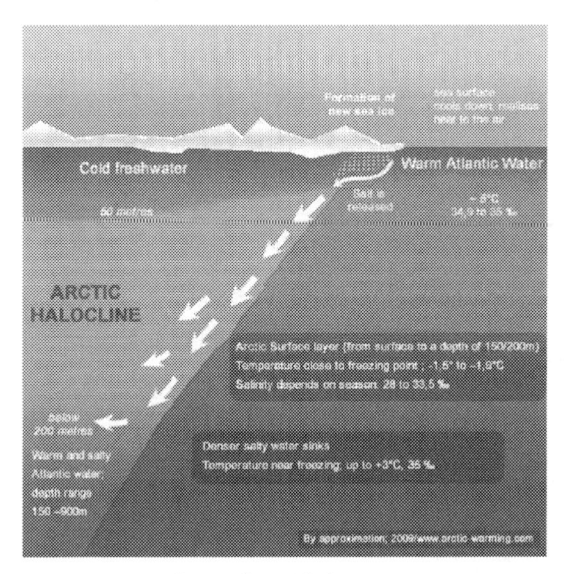

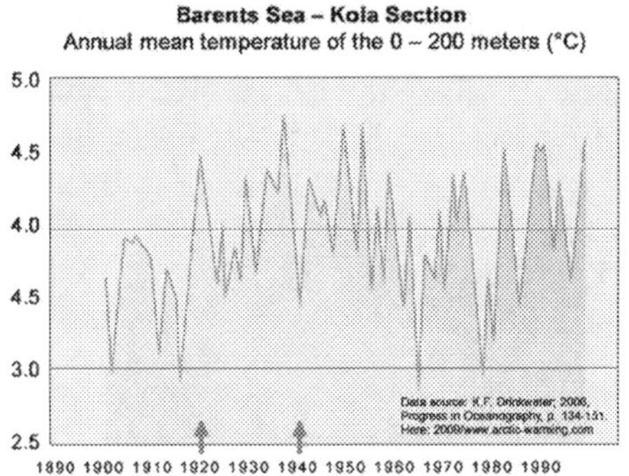

For the validity of the claim the authors (Bengtsson, 2004) should have answered these questions. Some may be not easy to verify, but others are. For example, had the strong warming impulse been generated in the Barents Sea realm, the temperature at Vardø-Norway should have jumped correspondingly to those in Spitsbergen. They didn't. Actually, the temperature record shows (see: Special Page "Questions to Bengtsson") that the temperature-level changed suddenly at about 1919, when the medium of the level prior 1919 was lifted by about one degree in the 1920s.

The optimum penetration of warm water at the 30 metre depth level occurs during the month of August. During winter this penetration is much less, and hardly capable of injecting substantial heat in the atmosphere. Actually some time ago the IPCC (IPCC, 1990) addressed this question quite differently, when saying:

Note

Norwegian & Spitsbergen Current

Extract from:
Berge, Jørgen, (2005) and Geir Johnsen, Frank Nilsen, Bjørn Gulliksen
„Ocean temperature oscillations causing the reappearance of blue mussels in Svalbard after 1,000 years of absence"
http://www.unis.no/40

These current systems transport Atlantic Water (AW), as an extension of the Gulf Stream system, with salinity >34.90 psu and temperature >3°C when it enters the Arctic Ocean. Here it meets the fresher and colder Arctic Water (ArW) with salinity between 34.3-34.8 psu and temperatures below 0°C. The transport time of water masses from Northern Norway (Lofoten / Vesterålen area at ~69°N, see Fig 1) to the shelf areas outside Isfjorden in Svalbard (a distance of approx. 1000 km) is estimated to be between 32-38 days for an average current speed between 0.30-0.35 ms^{-1} (Fig 1B-C). These estimates are based on time series observations of the NwASC and WSC. In terms of inter-annual variability, the annual mean volume transport of AW in the Svinøy section (a site at 62°N used for estimation of water transport by the NwAC) was at an absolute minimum in 2001 and increased to an absolute maximum in 2002.

>>The Norwegian Atlantic Current (NwAC); Norwegian Atlantic Slope Current (NwASC) & West Spitsbergen Current (WSC)<<.

Extract from:
Nilsen, J. Even Ø., (2006) and Eva Falck
"Variations of mixed layer properties in the Norwegian Sea for the period 1948–1999"
Progress In Oceanography, 70, p. 58-90

In the region around Ocean Weather Ship M (OWSM) in the Norwegian Atlantic Current (NwAC) the mixed layer depth varies between ~20 m in summer and ~300 m in winter. The depth of the wintertime mixing here is ultimately restrained by the interface between the Atlantic Water (AW) and the underlying water mass, and in general, the whole column of AW is found to be mixed during winter. In the Lofoten Basin the mean wintertime mixed layer reaches a depth of ~600 m, while the AW fills the basin to a mean depth of ~800 m. The temperature of the mixed layer at OWSM in general varies between 12 °C in summer and 6 °C in winter. Atmospheric heating controls the summer temperatures while the winter temperatures are governed by the advection of heat in the NwAC. Episodic lateral Ekman transports of coastal water facilitated by the shallow summer mixed layer is found important for the seasonal salinity cycle and freshening of the northward flowing AW. Atmospheric freshwater fluxes have no significant influence on the salinity of the AW in the area. (Position of Ocean Weather Ship M (OWSM): 66° North and 2° East)

Extract from:
Joanna Gyory, (2008) and A. J. Mariano, E. H. Ryan.
"The Norwegian & North Cape Currents."
www.oceancurrents.rsmas.miami.edu/atlantic/norwegian

The WSC is the northernmost extension of the Norwegian Atlantic Current. It flows poleward through eastern Fram Strait along the western coast of Spitsbergen. A mainly barotropic current, the WSC appears to be predominantly steered by the bathymetry. It is about 100 km wide and is confined over the continental slope, where it reaches its maximum current speed of 24 to 35 cm s^{-1} at the surface. Because it transports relatively warm (6 to 8°C) and salty (35.1 to 35.3) Atlantic Water, the WSC keeps this area free of ice. At around 79°N the WSC splits in two. The Svalbard branch stays close to the continental shelf of Spitsbergen, flowing north and east and eventually sinking and spreading at intermediate depths.

7. Where did the early Arctic Warming originate?

"Stronger westerlies over the Atlantic do not, therefore, account for the Arctic warming of the 1920s and 1930s on their own: in fact they preceded it by 20 years."

When nowadays researchers make an opposite claim, at least, they should discuss such conclusion, which the IPCC has made in a foregoing IPCC Assessment Report two decades ago. The authors should have also explained their statement that warming in the Arctic region reached its maximum during the period 1935 –1944 and that the strongest warming with more than 2°C occurred in the Kara- and Barents Seas (Bengtsson, 2004). As it stands now it implies that a warmer Barents Sea is responsible for this warming. At least it cannot verify the suddenness of the Arctic warming since 1919. For example:

- It takes about one year for a wave of warmth or of cold to traverse the distance from Kola meridian to Novaya Zemlya. (Schokalsky, 1936)
- The sea ice in the Kara Sea decreased only since about the mid 1920 (see Special Page).
- (Schokalsky, 1936, Notes): *"The other very important piece of work of the SADKO in 1935 consisted in the discovery of the edge of the continental shelf and of the deep furrow west of the Northern Land with saline and warm Atlantic water. This proves that the warm saline waters of the under –current of the Arctic Ocean penetrated to the deeper parts of the North Siberian shelf seas, which thus influenced the ice –condition of these shallow seas.*

The message from this explanation is clear; the warm water came from the interior of the Arctic Ocean, and not via the Barents Sea, which actually means that it entered the Arctic Ocean via the Fram Strait in the West of Spitsbergen. However, the warming of the Kara Sea was definitely later than the big temperature rise at Spitsbergen.

Although Dmitrenko et al. meanwhile published their recent findings:

> *We document through the analysis of 2002–2005 observational data the recent Atlantic Water (AW) warming along the Siberian continental margin due to several AW warm impulses that penetrated into the Arctic Ocean through Fram Strait in 1999–2000. The AW temperature record from our long-term monitoring site in the northern Laptev Sea shows several events of rapid AW temperature increase totalling 0.8°C in February–August 2004. We hypothesize the along-margin spreading of this warmer anomaly has disrupted the downstream thermal equilibrium of the late 1990s to earlier 2000s. (Dmitrenko, 2008);*

it is astonishing a bit that the early Arctic warming has never seriously been evaluated in conjunction with the warm Atlantic Water branch before or at the time it enters the Polar Sea.

d) Any other suggestions?

As far as this investigation is aware of, science seems to have not come up with more than the mentioned suggestions on how the natural commons of the Arctic-Atlantic fringe could generate and sustain this sudden warming.

We can therefore conclude this part of investigation, by stressing once again, that the sudden warming phenomenon was definitely not generated in the sea areas from the N-West, North and N-East of

7. Where did the early Arctic Warming originate?

Spitsbergen (80° N) for the simple reason that they had been permanently covered in sea ice. In winter, the cover could reach 70-100%, in summer, with great variations, around 50%. The sea ice 'shield' reduces to minimum all interaction between sea and air. Ice-covered seawater releases 90% less heat into the atmosphere than the open sea. It is possible that winter temperatures in the ice-covered areas could have increased very slowly over decades, but any assumption that such areas could have played a significant role in a dramatic temperature rise of the magnitude of Spitsbergen, is completely unfounded.

All areas situated north of 80° N and south of 66.5°N are excluded from the list of possible contributing factors. East Greenland Current is very cold and the Barents Sea depended completely on warm water supply from the South. Consequently, it is not difficult to confine the forcing of the Arctic warming since the winter of 1918/19 to a spot in the North Atlantic, namely Spitsbergen.

B. Warm Atlantic Water of Spitsbergen

After a lengthily discussion of all options as source for the early warming remains only the West Spitsbergen Current arriving with warm Atlantic Water at the Fram Strait, where it cools down and thereon submerges in the Arctic Basin. The conclusion is rectified by three further parameters, namely the facts:

- That nowhere else arrives so much heat from southerner regions so far North;
- That nowhere else the ocean surface remains usually all year round sea ice free so far in the North;
- That nowhere else in the Polar Regions the sun has no direct influence on the weather and climate over several winter months.

In summary it can be concluded, that nowhere else could such a sudden temperature jump have occurred, together with the observation of a huge difference between summer and winter temperature rise at the time of commencement and over a two decade period. But having nailed down the source of the warming does not say anything about the WHY. Why did the ocean current off Spitsbergen change course. As this shall be discussed in the next Chapter, a brief overview of the "Warm Atlantic Water" shall be given.

This sub-branch of the Atlantic Gulf Current crosses the Norwegian Sea in considerable depth northwards until reaching the high North in the Arctic realm as Spitsbergen Current. Once the shelf of Spitsbergen (ca. 77°N) has been reached, the current splits in two and passes the West and the East of Spitsbergen, to sink, eventually, into the Arctic Basis. The incoming water is relatively warm (6 to 8°C) and salty (35.1 to 35.3%) and has a mean speed of ca. 30 cm/sec^{-1}. From thereon the warm current goes through a series of highly complex processes. As no ocean observing systems were in place in the late 1910s, any theoretical analysis would hardly bring any relevant results because there are too many components involved in the transformation process of the warm Atlantic water into cold Arctic Ocean water.

At the sea surface, major components are air temperature, wind, waves, sea ice, ice motion and rain- or melt-water. Below the sea surface, there are only two components, which might represent overriding forces on ocean dynamics: seawater temperature and its degree of salinity. Density, the third major component, becomes a significant factor at greater depths.

7. Where did the early Arctic Warming originate?

While the water temperature and the salinity for internal oceanic dynamics is generating forces in every ocean water around the globe, the matter is particularly crucial with regard to the Spitsbergen Current. There is no other place as 'sensitive' as this one. Very warm and saline water arrives in a very cold environment. Nevertheless, the principal rules of ocean dynamics are simple:

- Warm water is lighter than cold water.
- Salty water is heavier than less saline water.

These two components allow uncountable variations and the sea areas around Spitsbergen have an increased range of variability.

Finally, we have to take into account the 'capacity' issue and the fact that the warming at Spitsbergen was the most pronounced during the winter. In winter, the importance of the ocean role for the supply of the atmosphere with heat becomes much more obvious. And here it comes in discussion the capacity issue. In average, a sea surface layer of mere three metres holds the same heat as an entire air column of 10,000 metres. One can explain it with a 'one-degree-image'. If 1° of heat is taken out of the upper three-metre of the sea surface layer, the entire atmosphere above warms up with one-degree. This is a relation which stresses out the importance of the transfer of the warm Atlantic water into the Polar region.

One needs only to pay attention to the interesting ice-cover charts for April 1918 and 1919 (see e.g. Chapter 2), which show that towards the end of the winter season the open sea area is reduced to a small percentage of about 10-20%. The section from were high winter temperatures could have only been released from an open sea area is the SW-sector of Spitsbergen, and that is the section where the West Spitsbergen Current transports the warm and saline Atlantic water towards the permanently ice-covered Arctic Basin.

The sudden warming at Spitsbergen after the winter of 1918/19 could have been caused only by a powerful heat resource force: the sea which, in this case, needed an additional forcing mechanism, namely either the warm Atlantic water or a big change in the 'dynamics' of the water body of the Nordic Sea. It could clearly be indicated that the sea areas around Spitsbergen in combination with the West Spitsbergen Current flowing into the Arctic Basin had been the sole driving force of the sudden Arctic warming in the early 20th century.

Chapter 8
Caused Naval War the Arctic Warming?

A. Which are the potential forces available?

Around the winter of 1918/19, nature had run its normal course. No "natural" event, as asserted by Johannessen (Johannessen, 2004), which could have affected the natural commons, had been observed in the North Atlantic or Arctic region, or at a global level. There was no significant earthquake, no eruption of a forceful volcano, no tsunami, no sunspots, and no big meteorite fell on the continent or into the sea. As previous analysis showed it, there was no hot spot in the atmosphere, from which warm air could have been transferred to Spitsbergen, causing a very pronounced warming and sustaining the phenomenon for such a long time. The only conclusion so far is that the sea areas around Spitsbergen must have undergone dramatic changes in a very sudden and unexpected manner. What must happen for science to regard something unexpectedly and out of tune with statistical average as worth to investigate, to understand, and to explain.

It is evident that the Spitsbergen event was, in the common sense of the word, 'unnatural'. Science has never recorded a similar situation again. To quote Bjerknes once again, this rise had been probably the greatest yet known on earth. As there was no extraordinary event in the space, in the atmosphere or in the common ocean behavior observed which might have caused this special phenomenon, it is reasonable to think about a causational force never experienced before: the First World War. Highly destructive forces had been fighting in the air, on land and at sea, in Europe, from August 1914 until November 1918, when the big warming at Spitsbergen began to manifest itself.

It is not so easy showing a pathway to a convincing solution, as to set up an interesting hypothesis. A correlation of naval war and weather has not yet been established, and any considered correlation is most likely not driven by mere shelling and bombing. The result of naval war activities in the natural commons, which is affecting the weather and climate, occurs in a chain of setting causations is a lengthily transformation process. The theorization in this respect does not assume in any way that naval war activities actually caused the Arctic warming during the First World War, but that naval war might have contributed to the event in whatever margin. If WWI contributed only 1% or less, it would require to be acknowledged in climatology. If the proportion were in any rate greater, it would be a sensation, and the general public and the politicians should know. Not knowing about this correlation would be gross negligence and could hardly be regarded as a convincing scientific commitment.

For demonstrating this correlation we have to add to the analysis already done in the foregoing chapters, with regard to the sudden temperature jump, the location Spitsbergen, the timing winter1918/19; three further complexes:

- Naval War activities in the North- and Baltic Sea, the Eastern North Atlantic from the English Channel to the North Cape and Archangels.

- Extreme cold winter conditions during the second half of WWI, with exceptional sea ice conditions in the Nordic Sea off Spitsbergen in spring 1917.
- The intensity of naval activities since autumn 1916 that culminated with the laying of an 80,000 sea mines barrage between Norway and Orkney Islands in Summer 1918.

Although each item is a story on its own, their interrelation is often quite evident. To start with, a brief assessment of the naval war situation will be given before picking up the climatically relevant correlation-issues. However, it shall be noted that fact presentation and consideration shall help to answer the question: Why did it come to the temperature explosion at Spitsbergen in winter 1918/19?

B. Naval War, a force to recon

WWI had destructive effects on men and on the environment, but nothing changed the commons of nature as much as the naval war did. This notion derives from understanding that the oceans, together with the sun, determine the status of the atmosphere on a short, medium or long term. The author of this paper has suggested and discussed this matter in a number of publications[45]. The impact of naval warfare on the ocean environment is in so far unique because it includes two principal aspects: one which is destructive to men, ships, and materials, and another one which is changing the temperature and salinity structure of the seas, where naval activities have taken place.

The second aspect is certainly not the only one, which might have had a significant impact on the interior of the seas in question, but it is, presumably, the most important one. Particularly sea surface layers of 50 metres depth and shallow seas (like the North Sea) are highly complex entities, always under permanent change due to season, wind, rain, river water, melt water, ice, and so on. Huge water masses in Western Europe seas were churned upside-down by naval war activities. The Norwegian Current transports these water masses northwards, to Spitsbergen. The temperature and salinity structure of the water had certainly changed its composition.

C. Forcing potential of naval war during WWI

Timing and ship losses. Although WWI started in August 1914, naval war began in earnest only two years later, when a series of new weapons were put in use: sea mines, depth charges, new sub-marines, and airplanes. By then naval warfare had reached a destruction stage to which no one might have thought of only two years earlier. The situation became dramatic when U-boats destroyed more ships than Britain could build in early 1917. In April 1917, the same total rate of the previous annual rate of 1916, ca. 850,000 tons, was destroyed by U-boats.

In April 1917, Britain together with the Allies lost 10 vessels every day. During the year of 1917, U-boats alone sank 6,200,000 tons, which means more than 3000 ships, and, during the war months of 1918, another 2,500,000 ship tonnage. The total loss of the Allies ship tonnage during WWI is of about 12,000,000 tons, namely 5,200 vessels. The total loss of the Allies together with the Axis naval vessels (battle ships, cruisers, destroyers, sub-marines, and other naval ships) amounted to 650, respectively 1,200,000 tons.

[45] Reference can be found at: http://www.whatisclimate.com/

8. Caused Naval War the Arctic Warming?

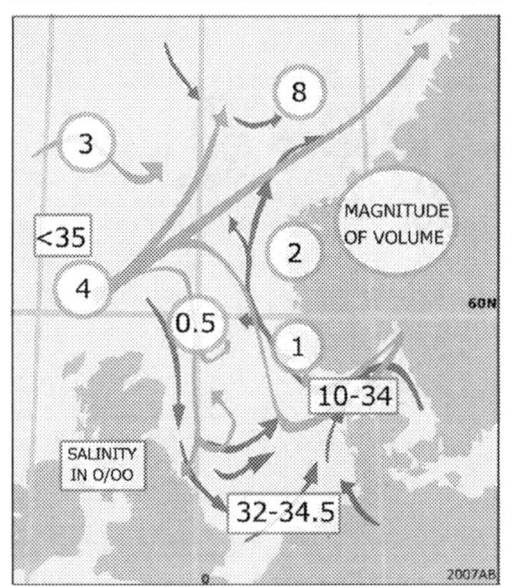

NORTH SEA CURRENTS

A weapon scenario churning the seas. The weapon scenario employed since 1916 is too complex to make a full assessment. Many figures are even impossible to quantify. The air force, for example, went through a great development. Airplanes were increasingly used in bombing and attacking missions over the sea. But it would be a mere speculation to try to indicate the number of bombs, which fell and exploded above or under the sea surface. We can say the same for the torpedoes activated or for the depth charges dropped upon the submarines, certainly many ten thousands of them. More detailed information is available about the sea mines. Sea mines were planted massively in the water column as soon as they became available since 1916.

A total of about 200,000 sea mines had been deployed. Of much powerful effect in churning the sea on a huge scale were those ships known under the name of minesweepers, which navigated the seas day and night to find and destroy the mines. Britain alone had more than 700 operational minesweepers; the Germans came close, too.

Churning the sea. War matters are usually quantified on the basis of costs and destruction caused to soldiers, population, buildings, industries, material, etc. Whether the water masses of a sea body have been turned upside down has never been of any interest. But that has happened on a grand scale. While in many cases seawater may have remained unchanged, temperature and salinity structure over a range of one meter to many dozen metres of surface water was always altered by any naval activity, whether there were weapons, sunken ships or mines planed or swept. Naval war at the magnitude of WWI means that many thousands of vessels navigated in defense-, combat-, or training missions, day and night. Battle ships had a draft of ten metres and could travel at a speed of 30 knots/hour (ca. 60 km/h).

In addition, the wide range of other impacts should be at least mentioned. Most ships that were sunk transported a variety of cargo, and all of them had equipment and provisions on board. The total number could be somewhere in the range of 10-15 millions tons. It has been never quantified how much cargo and provisions surfaced and traveled with the currents towards the Arctic region and how the sea and sea-ice interacted with all that stuff - a matter that should not be ignored outright.

Note

Weather protects impertinent attacker

Extract "Climate Change & Naval War"

Victoria/Canada, p.277 (references not shown);
And Chapter 5_15 at: http://www.seaclimate.com/

German battle cruiser bombards North Yorkshire's coast, 16 December 1914: The story is about weather-making by naval forces in combat missions at sea and is taken from the book 'Swept Channels' (Taffrail, 1938). The narrative tells the story of a German battle-cruiser bombarding Hartlepool, that had a battery of guns, and Whitby and Scarborough, that had not, shortly after daylight on December 16, 1914. That left 120 people killed, and over 400 wounded. A German Communiqué short time later reports about "parts of our naval forces", but does not name the vessels involved. It was claimed that one English cruiser was destroyed, others damaged. It follows the excerpt:

"The whole story is told by Mr. Winston Churchill in the *World Crises, 1911-1914, Vol. I, p. 467.* Squadrons and flotillas were moved to deal with the expected raid, and these forces actually made contact with the enemy during their retreat and opened fire. At one point the British and German battle-cruiser forces were only twenty-five miles apart, and were still closing in on each other. Further seaward there was a powerful battle squadron under the command of Sir George Warrender. The action was imminent, and it could only have one result.

Then, as it so often had happened before, the weather supervened. The wind sprang up and the sea started to run high. The North Sea mist came down until the horizon became blotted out in a curtain of thin vapor. The weather gradually thickened, the visibility dropping from 7,000 to 5,000 yards, then to 3,000. In the driving rain-squalls the area of vision was bounded by a circle whose radius was sometimes less than a mile.

Between fifteen and twenty heavy ships, and a number of light cruisers and destroyers, all steaming at high speed, were groping for each other within a space of about sixty square miles. Their wireless signals could be overheard in Whitehall, where their positions were constantly plotted on the large chart in the War Room at the Admiralty. It was like a nerve-racking game of Blind Man's Buff played in the dark, with huge ships instead of children – and the enemy escaped."

D. The connection between naval war and the Arctic warming

The naval war from1914 to 1918 can be considered as the most comprehensive single event in the late 1910s that has altered the common sea body structure around Great Britain through a huge variety of activities and means. In previous sections, we have proved that an extraordinary warming phenomenon took place at Spitsbergen. These two events are strongly connected by the timing of each event and bythe current system linking the two locations. No other coincidence of such a close relation has ever been observed before or after WWI. Is this a *prima facie* evidence that naval war could have contributed in causing the warming?

Correlation I – Extreme Sea Ice at Spitsbergen in Spring 1917

Here we have a story, which every one would call far-fetched. We too! Nevertheless, the story needs to be told – albeit here only in brief - as it may show what naval war is capable to do to the natural common. The brevity is due to the fact that naval war did presumably the same to the regional weather conditions as it did it during the Second World War (WWII), which is lengthily investigated elsewhere (Bernaerts, 2005)[46]. The point to make is two fold:

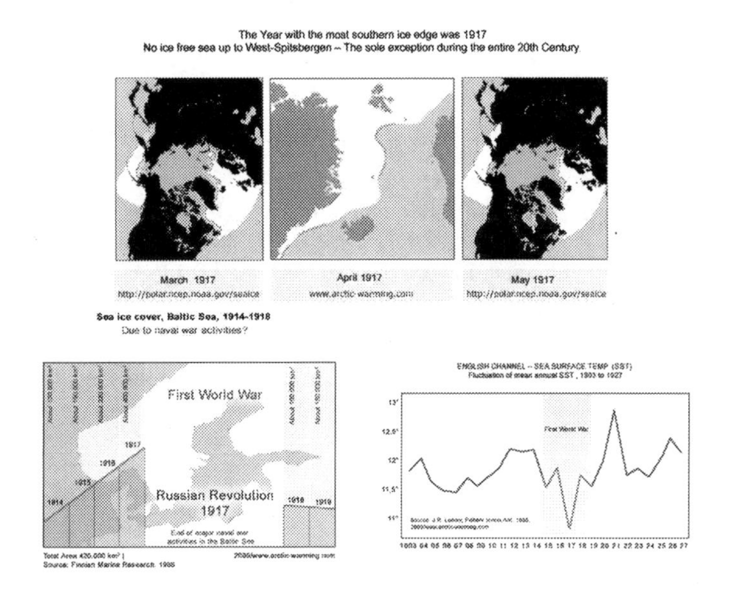

- The sea ice off Spitsbergen was extreme during spring 1917; actually it was the only time during the last century when the sea ice extended so far South by almost reaching the latitude of Bear Island. During all other years since 1900 a sea-ice free tongue up to the Fram Strait remained open sea. This lead to the question, whether the generation of higher salinity above the West Spitsbergen Current (WSC) during the freezing process, or the subsequent infusion of a huge fresh water component during the melting in summer 1917 initiated or promoted the big warming at Spitsbergen in less than 24 months later.

---

[46] See also: http://www.seaclimate.com/; http://www.warchangesclimate.com

8. Caused Naval War the Arctic Warming?

- The other aspect derives from naval war activities since August 1914 and can be again divided in two aspects, here formulated as questions:
    - Did naval war activities generate sea water conditions that supported the sea surface in the Spitsbergen region to freeze earlier, more heavily and extensively?
    - Are naval war activities partly responsible for the extreme cold winter conditions in North Europe and elsewhere in the Northern Hemisphere, through which the extensive freezing of Spitsbergen was enhanced?

Fact is that the war winters prior 1919 had been very cold in Europe. At the top stands the winter 1916/17. For Great Britain it was the third coldest on record[47]. After a cold series from 1881-1888 the coldest winters in the Arctic had been 1902, 1917, 1918, and 1966 (Kelly, 1982), and the coldest temperature measured on Spitsbergen was – 49.2°C measured on the 28th March 1917. One year later New York was served with the coldest in history of the city, wrote the New York Times in April 1919 remembering that it occurred "when Uncle Sam was feverishly trying to hurry his supplies to the East for shipment to Europe, when Jack Frost hampered transportation and coal was scare, are hard to forget".[48]

But a loose collection of some distinct weather conditions or significant deviation in statistics will prove little at this place. That does not say anything whether this aspect should not receive more attention, bearing in mind that the naval war got very serious only in autumn 1916 and remained devastating over two winters until 1918. Three examples may indicate the possible relevance:

__The Baltic Sea was an extensive scope for all sort of naval activities from August 1914 until the Russian October Revolution 1917. The sea ice cover increased with the length and intensity of the fight, as illustrated by the attached image, but diminished since winter 1917/18, when the naval war in the Baltic Sea ceased due to the Russian October Revolution. Since winter 1917/18, the Baltic Sea had been left to a rest, fact immediately reflected in a much lower sea ice extension. Had fighting continued, had this sea got another record cold year?

__The Royal Navy had to do a lot of surveillance, mine sweeping and war activities in the English Channel. For example in September 1916 a flotilla of about 570 anti-submarine vessels were on hunt for three U-Boats operating for about one week between Beachy Head and Eddy Stone Light. The U-Boats were able to sink thirty ships without having been unscathed themselves. The operation of such a flotilla is presumably reflected in the record of sea water temperature (SST) taken in the English Channel for the year 1916-1917 (see image). This suggests the conclusion that a cooling of the English Channel water by naval activities will inevitably support cold winter conditions in Great Britain including Ireland and southern part of the UK.

__The third example is an observation made by a scientist from Kew Observatory near London in 1942: "Since comparable records began in 1871, the only other winters as snowy as the recent three (1939-1942), were those of the last war, namely 1915/16, 1916/17, and 1917/18." (Drummond, 1942)

---

[47] Web page: t.a.harley; http://www.personal.dundee.ac.uk/~taharley/1917_weather.htm

[48] Pearson, Samuel K. (1919); "Illusion about Weather!, The New York Times, the 6th of April 1919.

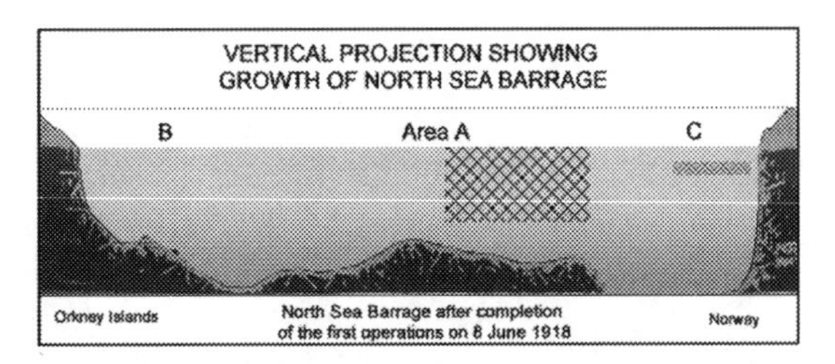

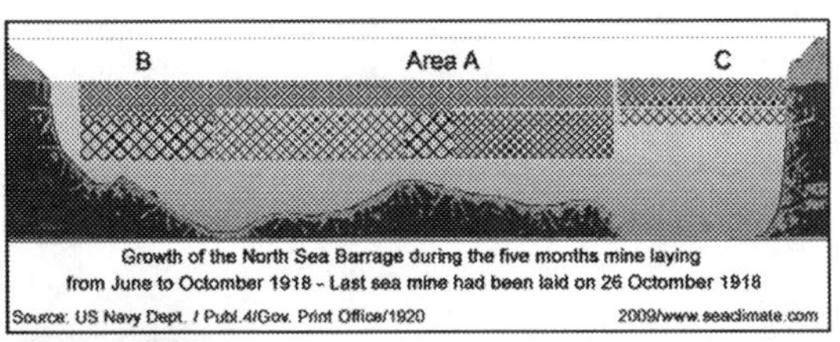

Whether or whether not the naval war contributed to the harsh winter conditions, for ocean research it would be still high time to show that they analyzed and understand the consequences of extraordinary sea ice conditions between Bear Island and Spitsbergen in winter 1916/17.

Correlation II – The sea water from West Europe ends up at Spitsbergen

Even the wildest guess will hardly tell with any precision how intensive the seas around Britain have been churned, mixed, or altered by naval activities during WWI. The hazard to the marine environment will not be coming from any sort of pollution but from the mere mixing of sea water structures, concerning temperature and salinity, over many meters depths below the sea surface. Thus the complete sea water body of the North Sea could have been turned over several times during the war. Assuming that this has no effect on the 'natural common' could be called naïve. At least during a number of months the temperature structure cover considerable ranges according to the sea level. The same applies for the salinity, which is further complicated by rain, river run-off, etc. For example, any rain water would usually "swim" on a body of saltier water until it is either much colder than the sub-water layer, or the wind would start mixing the sea surface layer. Naval war activities would add a third option. The sea structure has not remained any longer the structure the sea used to have. The changed structure is transferred by the ocean current system to other regions.

8. Caused Naval War the Arctic Warming?

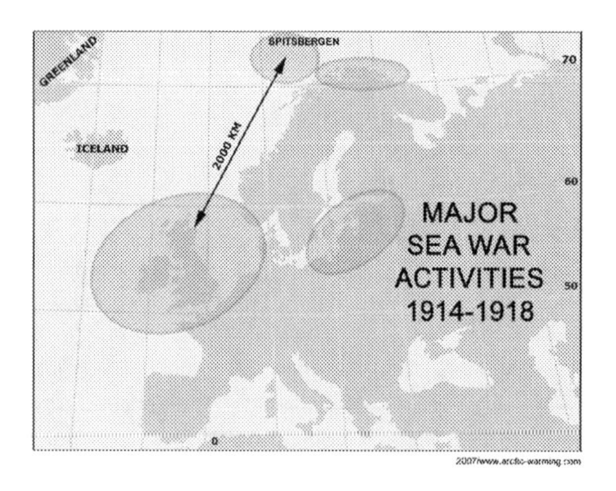

All naval activities around Britain had changed the water structure that moved on toward the North. These activities culminated in building the monstrous Northern Mine Barrage at the northern out-let of the North Sea between Orkney Island and Norway from spring to October 1918. A total of 80,000 sea mines were laid. The aim was to prevent U-Boats from leaving the North Sea. For further details see the Special Page. The sea area used for the Barrage is complex, as the attached image indicate. The area consists not only of brackish North Sea water, but warm Atlantic water is flowing around the Orkney Islands and enters the North Sea. The outflow from the North Sea is along the Norwegian Coast up to the North.

NORTHERN BARRAGE
70'000 sea mines laid from June to Octomber 1918 to stop U-Boats

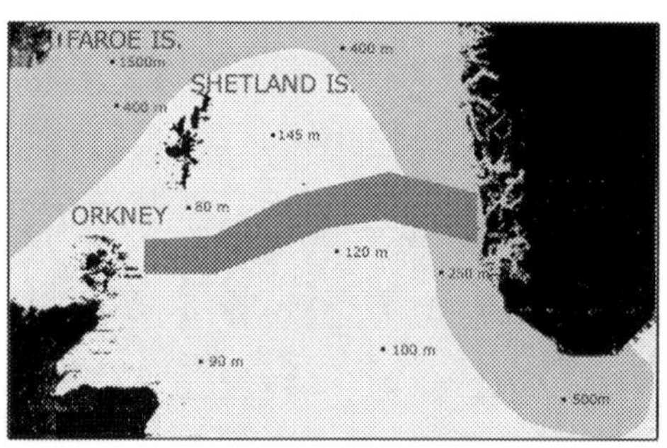

Source: US Navy Dep., Publ. No. 2&4, 1920.

How close was the naval war to Spitsbergen?

The distance between Spitsbergen and the main naval battleground was of about 2000 km. But this distance is not very significant in this case. The currents moving along the Norwegian coast consist of water from the North Sea and of water from the Golf Current, flowing at a medium speed of 0.1 km/hour. At the sea surface, the current is up to 10 times faster.

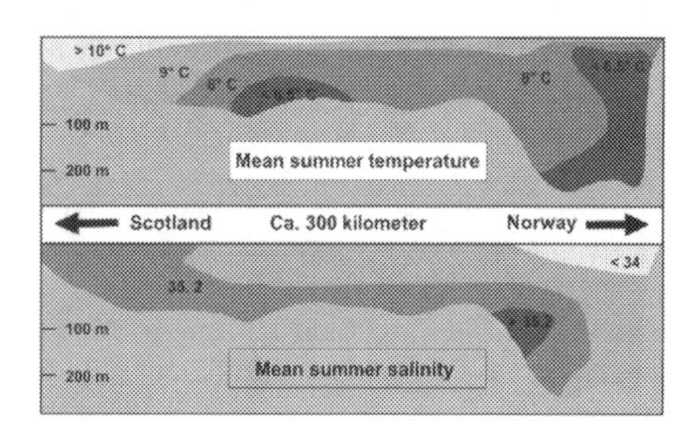

- The branch of the North Atlantic Current has temperatures exceeding 6°C and salinity greater than 35. The main arm is well below the sea surface and in quite a distant to the coast of Norway.
- Norwegian Coastal Current flows closer to the coast of Norway in the upper 50-100 m of the water column with lower temperatures than the Atlantic branch and low-salinity water, less than 34.8.

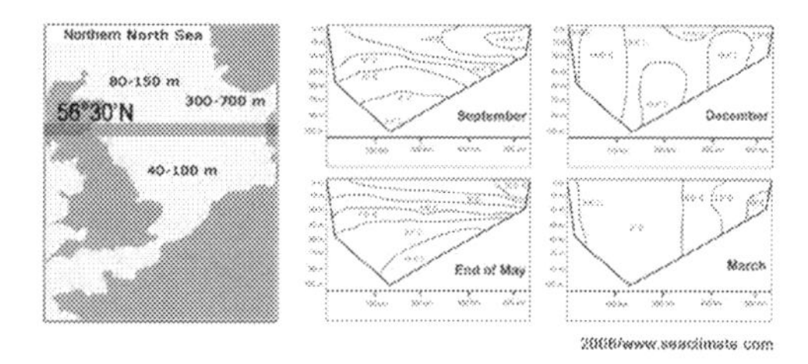

The current speed between the two currents is not equally high, but the distance of ca. 1000 km from the Lofoten to Spitsbergen can be covered by 4 to six weeks (Berge, 2005), although the Atlantic branch requires more time. While the Atlantic branch current needs some time to cover the distance between Scotland-Shetland Islands and Spitsbergen (ca. 1500km), the transport of surface water into the high North can be accomplished within few weeks or months. All

1920

Daniels, J.

'The Northern Barrage'

No. 2 and No..4 - Publication, Navy Department, Washington 1920.

SUMMARY OF THE BOOKS

U-boats had been a serious threat to the Allies since 1916. They regarded it paramount to prevent U-boats from leaving the North Sea into the Atlantic. To 'close' the northern outlet of the North Sea, about 150 sea miles (ca. 275 km), a long barrage between the Orkney Islands and Norway would be required. Off the Norwegian coast the water is 300 metres deep and the coast off Orkney about 100 metres. Sea currents can reach 3-4 nautical miles/hour. That was a challenge and required the development of a new mine, the MK6, to meet it. The charge consisted of 300 pounds of grade B trinitrotoluol (TNT). The mine itself was supposed to have a destructive radius of 100 feet (ca. 30 m) against submarines. Calculations showed that approximately 100,000 mines should effectively prevent U-boats from passing the line. Actually, only about 70.000 mines were laid until October 1918.

Mines were available by March 1918, laying started. "Shortly after mine laying had commenced mines began to explode prematurely. By counting the explosions it was estimated that between 3 and 4 per cent of 3,385 mines laid blew up". In the middle section "A" mines were supposed to be laid as follows: 10 rows of mines at 80 feet submergence, 4 rows of mines at 160 feet submergence, 4 rows of mines at 240 feet submergence. Corresponding rows were laid before the Orkney Islands (section B) and Norway (section C).

From a detailed account by Daniels here are some illustrative events:

__When deep level mines exploded, 'a circle of brown discolored water was spreading slowly around the vessel'. July 6th, a mine had been found on the Norwegian coast in the vicinity of Bergen (Daniels, p.108).
__July 14th, 5,395 mines had been laid in 4 hours and 22 minutes (Daniels, p.109).
__Approximately 5% of the mines exploded prematurely – a slight increase over previous statistics.
__July 29th, 5,399 mines laid with 14% of mines going off (Daniels, p.111), at one time even 19% in section C (Daniels, p.112).
__August 18th, 12% of mines exploded prematurely.
__Section A; mines which had been laid in this area by the British in March 1918, had in the meantime been swept up. (Daniels, p.115).
__September 29th, the Norwegian Government said that mines would be laid in the vicinity of Udsire Island, and it is understood that this had been done by October 07th (Daniels, p. 119).
__With the signing of the armistice on November 11th, the building of the mine barrage ended. (Daniels, p.120).
__Final Status of Barrage (extract): up to November 11th a total of 56,760 United States and 16,300 British mines have been laid. Completion of the barrage within the Norwegian territorial waters had been effected by Norway herself.

Mine sweeping started in spring and ended in autumn 1919. From more than 73.000 mines

__about 5,000 exploded prematurely soon after laying
__20,000 mines were disposed of while the work was in progress
__from the remaining ca. 50,000 mines
__more than 30,000 mines were already 'gone' in spring 1919, either drifted away, or exploded during winter storms;
__rest 20,000 were swept in 1919.
Six months of sweeping operation comprised seven sweeping missions involving more than 70 vessels and 10 supply vessels.

mentioned timing, although only rough estimates, illustrates perfectly the "connection" between WWI and Spitsbergen warming.

E. The system shift

What does a system shift mean in respect to the Spitsbergen region? The main answer is simple. The incoming warm water of the West Spitsbergen Current was "positioned" in a manner that it could release more heat into the atmosphere. This can happen in two ways: (1) the sea ice forming during the winter season diminishes, which would not explain the suddenness of the shift; or (2) the thickness of cold sea water layer above the warm water was suddenly substantially reduced so that the air temperatures could immediately benefit from warmer water close to the sea surface. This was actually the case. In the mid 1930s it had been already discovered and published, that since the FRAM expedition in 1893-1896 the cold surface layer had grossly weaken:

> *"The branch of the North Atlantic Current which enters it by way of the edge of the continental shelf around Spitsbergen has evidently been increased in volume, and has introduced a body of warm water so great, that the surface layer of cold water which was 200 metres tick in Nansen's time, has now been reduced to less than 100 metres in thickness. "* (Schokalsky, 1936)

Between the two observations, by Nansen and Russian research vessels, lays a time span of more than three decades. Was the diminishing process gradually over the full time period? Definitely not. To be clear, it is the observation of a sudden shift. The temperature could not 'explode' during the winter season without more heat release from the warm water current which could only be sustained by a thinner cold surface layer, respectively by more war water from the South. That was a sudden system shift. Science should have been able to explain the findings since long. Brooks noted 70 years ago: "Whatever the mechanism, the rise of temperature did begin and presumably had a cause". (Brooks, 1938)

F. Summing up

The chapter covered a quest for the reasons of the Arctic warming since winter 1918/19. The goal was high; and achieved? The judge is out. On one hand one could say: Correlation is not causation. But science without investigating a surprising correlation would get not far. Correlation is the spice to take actions and to ask: Why. We could demonstrate a case where the correlation of events, magnitude, space, location, and timing, are so closely intertwined that one is forced to assume that the connection is inevitably the source of causation. That is called a prima facie evidence. And a prima facie evidence between the naval war activities in West Europe waters and the system shift to the West Spitsbergen Current in the sea area off Spitsbergen could be established. That is not necessarily a 100% proof, but it is enough to require from every claim, otherwise dissembling the demonstrated strong correlation, and establishing another, more solid, evidential conclusion. As there can be no doubt that only the warm Atlantic water in the West Spitsbergen Current could have initiated the big Spitsbergen warming in the late 1910s, and sustain it in the region for two decades, the naval war thesis can only be challenged with conclusive evidence that the dramatic system shift in the Northern north Atlantic stand in correlation with another event, respectively the system shift in the current was 'natural'. But due to the strong correlation with WWI, this cannot only be merely claimed, but the claimant should establish such claim on solid proof.

Chapter 9
CONCLUSION

To many climate scientists the Arctic warming remains "*one of the most puzzling climate anomalies of the 20th century*" (Bengtsson, 2004). Yet, the phenomenon discussed here is not as puzzling as claimed. This investigation could establish that only the seas in the realm of Spitsbergen could have generated the sudden increase of the observed air-temperatures, and indicate the precise time period, namely the winter of 1918/19. This timing stands in extremely close relation with the naval war activities in Europe.

Ninety years have passed since the most pronounce and sudden climatic shift has occurred. The 'when' and 'where' had been the main task of the investigation. Although the scientists of the pre-WWII generation started only 10 years after the Arctic warming commenced, their one decade long elaborations was in no way less to the point than much more recent investigation. If WWII had not ended this attention, the 1930s generation might have explained the early Arctic warming since long.

It came differently. On the warming period followed an almost four decade long cooling. In so far it could serve as excuse for showing little interest in the pre WWII warming. But the trend altered direction again. At the change of the millennium the Arctic temperatures were as high as in the late 1930s. Latest when the warming returned in the 1980s it was high time to investigate and explain the previous warming since the late 1910s. That was 30 years ago. But nothing has been done. Many hundred papers were published, but the "climatic revolution" (Ahlmann, 1946) and what made it happen had not been regarded as worth to receive the required attention.

Instead of explaining the first warming that happened under the eyes and observations of modern science, the issue is pushed aside by claiming "natural variability". That is a non explanation. It generates a wrong impression. If a hurricane destroys New Orleans, it was a hurricane that destroyed the city and not "natural variability". If a tsunami sinks dozen of ships, a tsunami sank the ships. If the West Spitsbergen Current warmed the Arctic, than it was a branch of the Gulf Current that increased the Arctic temperatures. It was therefore necessary to establish to the point, that the warming started at Spitsbergen in winter 1918/19, that this even affected the temperatures in Greenland from ca. 1920 to 1933, and in the East of Spitsbergen the warming lasted until the early 1940s. Concerning Europe there was a warming over two decades from ca. 1920 to 1940, but this warming was presumably not generated alone from the Arctic region, but has had a regional or continental component as well. Current Arctic research should understand what had caused the Big Spitsbergen warming early last century.

The investigation could furthermore demonstrate that there is a high possibility of a connection between the Arctic warming and the naval war in Europe from 1914 to 1918, due to the fact that the seawater current system had carried all the war torn sea water literally into the front garden of the Spitsbergen. Had the naval war of WWI occurred in the sea area of Spitsbergen at a similar magnitude as around Great Britain during the war, presumably no one would have ever questioned the interconnection between the Arctic Warming and the naval war.

Acknowledgement

This book would not have been possible without the invaluable help and commitment from Angela Boartes. The liaison commenced in 2006 and includes the care for the websites www.whatisclimate.com and www.arctic-warming.com. Arctic warming became the first big challenge by preparing a conference paper for PACON 2007/Honolulu and a website, and Angela supported it greatly. When it was decided to present the complex matter in great detail and underline the text with explanatory images, Angela and her team showed high professionalism and willingness. It made the research, the elaborating of material, and the writing a smooth and enjoyable undertaking. Angela also undertook all the formatting and publication procedure. With appreciation I express my very cordial thanks.

Arnd Bernaerts/February 2009

Bibliography

Ahlmann, H.W. (1946); "Research on Snow and Ice, 1918-1940", The Geographical Journal, 1946, p.11-25.

Ahlmann, H.W.(1953) "Glacier Variations and Climatic Fluctuations". Series Three, Bowman Lecture Series, The American Geographical Society, George Grady Press, New York ; at: http://www.questia.com/PM.qst?a=o&d=1918470

Aagaard, Knut, (1982), in: Louis Rey, The Arctic Ocean, „The Climate Environment of the Arctic Ocean", Comite Arctique International, p. 69-81.

Berge1, Jørgen, (2005) and Geir Johnsen, Frank Nilsen, Bjørn Gulliksen; „Ocean temperature oscillations causing the reappearance of blue mussels in Svalbard after 1,000 years of absence"; http://www.unis.no/40;

Bengtsson, Lennart (2004), Vladimir A. Semenov, Ola M. Johannessen, The Early Twentieth-Century Warming in the Arctic-A Possible Mechanism, Journal of Climate, October 2004, page 4045-4057.

Bernaerts, Arnd (2005); "Climate Change & Naval War – A Scientific Assessment-", Victoria/Canada, pages 326. Ditto: http://www.seaclimate.com/

Birkeland, B.J. (1930) , Temperaturvariationen auf Spitzbergen, Meteorologische Zeitschrift, Juni 1930, p. 234-236

Bjerkness, J. (1959); 'The Recent Warming of the North Atlantic'; in: Bolin, Bert, ,The Atmosphere and Sea in Motion', Oxford 1959, p. 65ff.

Brönnimann, S. et al ed;(2008/09), Brönnimann; Luterbacher, J.;Ewen,T.;Diaz, H.F.; Stolarski, R.S.; Neu, U.: "Introductory Paper" in: A Focus on Climate During the Past 100 Years; Dordrecht, p. 364.

Brooks, C.E.P. (1938), "The Warming Arctic", The Meteorological Magazine, p.29-32.

Carruthers, J.N. (1941), "Some Interrelationships of Meteorology and Oceanography"; Quarterly Journal of the Royal Meteorology Society, p.207-232.

Covey, Curt (1991); "Chaos in ocean heat transport", Nature, 353, Oct.1991; p.796-797.

Chylek, Petr; (2006), and M. K. Dubey, G. Lesins; "Greenland warming of 1920-1930 and 1995-2005", Geophysical Research Letters, 33, 13 June 2006, L11707.

Daniels, Josephus; (1920); 'The Northern Barrage', No.2 (The Northern Barrage and other mining activities), No.4, The Northern Barrage –Taking up the mines), Publication, Navy Department, Washington Government Printing Office, 1920.

Delworth, Thomas L.; (1997), & Syukuro Manabe and Ronald J. Stouffer; „Multidecadal climate variability in the Greenland Sea and surrounding regions: a coupled model simulation"; Geophysical Research Letters, Vol. 24, NO. 3, p. 257-260.

Dmitrenko, I. A, et al. (2008) ; Toward a warmer Arctic Ocean: Spreading of the early 21st century Atlantic Water warm anomaly along the Eurasian Basin margins", J. Geophys. Res., 113, C05023, doi:10.1029/ 2007JC004158. Co-authors: V. Polyakov, S. A. Kirillov, L. A. Timokhov, I. E. Frolov, V. T. Sokolov, H. L. Simmons, V. V. Ivanov, and D. Walsh.

Drinkwater, K.F. (2006), "The regime shift of the 1920s and 1930s in the North Atlantic", in: Progress in Oceanography, p.134-151.

Eythorsson, J. (1949) 'Temperature Variation in Iceland", Geografiska Annaler, p.36-55.

Gore, Al (Albert), (1992); "The Earth in Balance", London, 1992.

Gore, Albert (2007); Al Gore: Moving Beyond Kyoto, The New York Times, July 1, 2007, WK 13.

Bibliography

Gyory, Joanna (2008) and A. J. Mariano, E. H. Ryan. "The Norwegian & North Cape Currents."; www.oceancurrents.rsmas.miami.edu/atlantic/norwegian

Hesselberg, Th. (1956); et al: Johannessen, T.W.; in: R.C. Sutcliffe (ed), Polar Atmosphere Symposium (Oslo 1956), "The Recent Variations of the Climate at the Norwegian Arctic Sectors", London, 1958, p18-29.

Henning, R.; 'Die Erwärmung der Arktis', in: WuK, No. 1-2, 1949, p. 49-51.

IPCC (1990); by J.T. Houghton, G.J. Jenkins, J.J. Ephraums (ed); Climate Change –The IPCC Scientific Assessment, Chapter 7, Observed Climate Variation and Change, Cambridge 1990,

IPCC (2007) - Climate Change, WG I ; The Physical Science Basis; Summary for Policymakers; released in Paris on February, 2nd, 2007, http://ipcc-wg1.ucar.edu/wg1/docs/WG1AR4_SPM_PlenaryApproved.pdf

Johannessen, Ola M. (2004); et.al: Lennart Bengtsson, Martin W. Miles, Svetlana I. Kuzmina, Vladimir A. Semenov, Genrikh V. Alekseev, Andrei P. Nagurnyi, Victor F. Zakharov, Leonid Bobylev, Lasse H. Pettersson, Klaus Hasselmann and Howard P. Cattle; Arctic climate change – Observed and modeled temperature and sea ice variability; Nansen Environmental and Remote Sensing Center, Report No. 218, Bergen 2002; Tellus 56A (2004), p. 328 –341, Corr. 559-560.

Johannsson, O.V. (1936); 'Die Temperaturverhältnisse Spitzbergens (Svalbard)', in: Annalen der Hydrographischen Meteorologie, Maerz 1936, pp. 81-96.

Jones, P.D. (1986/Feb.), Wigley and Kelly; 'Variations in Surface Temperatures: Part 1. Northern Hemisphere, 1881–1980', in: Monthly Weather Review, Vol. 110, February 1982, pp. 59-70.

Jones, P.D., (1986/July) T.M.L. Wigley, P.B. Wright; 'Global temperature variations between 1861 and 1984', in: Nature, Vol. 322, July 1986, p. 430-434.

Kahl, Jonathan D. (1993); et al: Donna J. Charlevoix, Nina A. Zaftseva, Russell C. Schnell, and Mark C. Serreze; "Absence of evidence for greenhouse warming over the Arctic Ocean in the past 40 years"; Nature, 1993, 361, p. 335 - 337

Kelly, P.M. (1982); et al: Sear, Cherry and Tavakol; 'Variations in Surface Air Temperatures: Part 2. Arctic Region, 1818-1980'; in: Monthly Weather Review, Vol. 110, p. 71-83.

Kirch, Regina (1966); ‚Temperaturverhältnisse in der Arktis währen der letzten 50 Jahre), Meteorologische Abhandlungen, Bd. LXIX, Heft 3.

Lamb, H.H. (1977); ‚Climate – Present, Past and Future', Vol. 2, London, 1st ed. 1977, 2nd ed. ca. 1980s, p.528.

Lamp, H.H. (1982); in: Louis Rey, The Arctic Ocean, „The Climate Environment of the Arctic Ocean", Comite Arctique International, p. 148, Fig 7.10(a to d)

Lee, A.J. (1955); "Influence of Hydrograph on the Bear Island Cod Fishery", in: Ministry of Agriculture and Fisheries", Fishery Investigations, Series II, Vol. XVIII, No.4, London 1955, p.72-107.

Manley, Gordon (1944); ‚Some recent contributions to the study of climatic change', in: Quarterly Journal of Met. Soc., Vol. 73, p. 197-219.

Milloy, Steve (2005); 'Arctic Warming Update', www.Junkscience.com, January 15, 2005.

Nilsen, J. Even Ø., (2006) and Eva Falck; "Variations of mixed layer properties in the Norwegian Sea for the period 1948–1999"; Progress In Oceanography, 70, p. 58-90.

NSDIC ; National Snow and Ice Data Center; http://nsidc.org/arcticseaicenews/index.html

Overland, J.E. (2004); et al: Michael C. Spillane, Donald B. Percival, Harold O. Mofjeld; "Seasonal and Regional Variation of Pan-Arctic Surface Air Temperature Over the Instrumental Record", Journal of Climate, Vol.17, pp.274-288.

Overland, J.E. (2005); and Muyin Wang: "The tird Arctic climate pattern: 1930s and early 2000s"; Geophysical Research Letters, Vol. 32, L23808

Overland, J.E. (2006), "Arctic change: multiple observations and recent understanding", Weather, Vol. 61, p. 78-83.

Overland, J . E. (2008); M. WANG & S. SALO; „The recent Arctic warm period", Tellus, 2008, 60A, p.589–597

Bibliography

Overpeck, J. (1997); and K. Hughen, D.Hardy, R. Bradley, R. Case, M. Douglas, B. Finney, K. Gajewski, G. Jacoby, A. Jennings, S. Lamoureux, A. Lasca, G. MacDonald, J. Moore, M. Retelle, S. Smith, A. Wolfe, G. Zielinski; "Arctic Environmental Change of the Last Four Centuries"; Science , 1997, Vol. 278; p. 1251 – 1256:

Palmer, Tim; (1991); in: Nina Hall, ed., Guide to Chaos; "A weather eye on unpredictability"; London, p. 69-81

Pohjola, V.(2007); „Arctic Warming – a Perspective from Svalbard", Global Change NewsLetter, No. 69, p. 9 -12

Polyakov, I. V., (2002) and G. V. Alekseev, R. V. Bekryaev, U. Bhatt, R. L. Colony, M. A. Johnson, V. P. Karklin, A. P. Makshtas, D. Walsh, A. V. Yulin; „Observationally based assessment of polar amplification of global warming"; Geophys. Res. Lett., 29(18), 1878.

Polyakov, I.V. (2003), et al: Roman V. Bekryaev, Genrikh V. Alekseev, Uma Bhatt, Roger L. Colony, Mark A. Johnson, Alexander P. Makshtas, and David Walsh; Variability and trends of air temperature and pressure in the maritime Arctic, 1875 – 2000; J. Climate, 16 (12), 2067-2077, 2003. Extract via: Website of the Int. Arctic Research Centre, IARC, Alaska; or: http://www.frontier.iarc.uaf.edu/%7Eigor/index.php

Polyakov, I.V. (2004); et al: Alekseev, L. A. Timokhov, U. S. Bhatt, R. L. Colony, H. L. Simmons, D. Walsh, J. E. Walsh, V. F. Zakharov, 2004, Variability of the Intermediate Atlantic Water of the Arctic Ocean over the last 100 Years, Journal of Climate, Vol.17, No.23, 2004.

Polyakov, I.V. (2005), et al.; „One more step toward a warmer Arctic"; Geophysical Research Letters, Vol. 32, L17605;
Co-authors: Agnieszka Beszczynska, Eddy C. Carmack, Igor A. Dmitrenko, Eberhard Fahrbach, Ivan E. Frolov, Rüdiger Gerdes, Edmond Hansen, Jürgen Holfort, Vladimir V. Ivanov, Mark A. Johnson, Michael Karcher, Frank Kauker, James Morison, Kjell A. Orvik, Ursula Schauer, Harper L. Simmons, Øystein Skagseth, Vladimir T. Sokolov, M. Steele, Leonid A. Timokhov, David Walsh, and John E. Walsh.

Sandström, J.W.; (1946-1947), „World Temperatures Anomalies", Kungl. Svenska Vetenskapsakademies Handlingar, Tredje Serien, Band 23, No. 4, Stockholm.

Serreze, Mark C (2006) and Jennifer A. Francis; "The Arctic Amplification Debate"; Climatic Change, 76, p. 241-264

Scherhag, R. (1936/März) 'Eine bemerkenswerte Klimaveränderung über Nordeuropa'; Annalen der Hydrographischen Meteorologie, pp. 96-100.

Scherhag, R. (1936/Sept.); "Die Zunahme der atmosphärischen Zirkulation in den letzten 25 Jahren"; Annalen der Hydrographie und Maritimen Meteorologie, p. 397-407, Fig 7

Scherhag, R. (1937); ‚Die Erwärmung der Arktis', in: Cons. Intern. Expl. Mer. Rap. Proc.- Verb., Copenhagen, Vol. 12, p. 263-276.

Scherhag, R (1939/Feb) „Die Erwärmung des Polargebiets"; Annalen der Hydrographie, LXVII. Jahrg.(1939), p.57-67

Scherhag, R. (1939/Juni).; 'Die gegenwärtige Milderung der Winter und ihre Ursachen', in: Annalen der Hydrographie und Maritimen Meteorologie, Juni 1939, pp. 292-302.

Schokalsky, J. (1936); ‚Recent Russian researches in the Arctic Sea and the in mountains of Central Asia', in: The Scottish Geographical Magazine, Vol. 52, No.2, March 1936, p. 73-84.

Serreze, M.C. (2000) and J. E. Walsh, F. S. Chapin III, T. Osterkamp, and other; Observational Evidence of recent change in the Northern High-Latitude Environment", Climatic Change, 46, p. 159–207.

Serreze, Mark C. (2006); et al. Jennifer A.Francis; „The Arctic Amplification Debate"; Climatic Change, 2006, 76(3-4): p. 241-264,

Stein, M. (2007), Warming Period off Greenland during 1800-2005: Their potential Influence on the Abundance of Cod and Haddock in Greenland Waters; J.Northw.Atl.Fish.Sci.; Vol.39; p. 1-20

Sverdrup, H.U. (1942); "Oceanography for Meteorologists", New York, 1942, Chapter X,

Taffrail; (1938) Captain Taprell Dorling; ‚Swept Channels', London (3rd ed).

Teng, Haiyan (2006, and Warren M. Washington, Gerald A. Meehl, Lawrence E. Buja, Gary W. Strand; "Twenty-first century Arctic climate change in the CCSM3 IPCC scenario simulations", Climate Dynamics, 26, p. 601 –616.

Bibliography

Tomczak, Matthias (2003), and J Stuart Godfrey: "Regional Oceanography: an Introduction" 2nd Ed. (2003), online: http://www.es.flinders.edu.au/~mattom/regoc/pdfversion.html
Wagner, A. (1940); 'Klimaänderungen und Klimaschwankungen', Braunschweig, 1940, p. 50.
Wang, Muyin (2004), and James E. Overland; „Detecting Arctic Climate Change using Köppen Climate Classification", Climate Change , Vol. 67, p. 43 – 62.
Weaver, A.J., (1991) & E.S. Sarachik and J. Marotzke; "Freshwater Flux Forcing of Decadal and Interdecadal Oceanic Variability", Nature, 353, p. 836-838.
Weikmann, L. (1942); ‚Die Erwärmung der Arktis', Berlin, 1942.
Xoplaki, E. (2006); "August 2006 is Warmest of Over More Than Half a Millennium", at: www.scitizen.com, Dec.23, 2006.
Zakharov, V.F. (1997) 'Sea Ice in the Climate System', Arctic Climate System Study, WMO/TD-No. 782, p.70f.

List of images

NOTE: The reference pages below apply for this b/w version
but not necessarily for the color version:
„How Spitsbergen Heats the World - The Arctic Warming 1919-1939"
Book on Demand GmbH, ISBN 978-3-8370-9524-1
and
The website color edition at: www.arctic-heats-up.com

List of Images

List of Images

List of Images

List of SPECIAL PAGES

Only lead author and abbreviated title indicated